MEDICINE
from
CAVE DWELLERS
to
MILLENNIALS

JONATHAN L. STOLZ, M.D.

To Joan and Ron —
With best wishes
for continued good
health

MEDICINE FROM
CAVE DWELLERS TO
MILLENNIALS

by

Jonathan L. Stolz, M.D.

TELEMACHUS PRESS

MEDICINE FROM CAVE DWELLERS TO MILLENNIALS

Cover designed by Telemachus Press, LLC

Cover art:
Copyright © iStockPhoto/871071872/WinnieVinzence
Copyright © iStockPhoto/529774419/d-l-b

Published by Telemachus Press, LLC
7652 Sawmill Road
Suite 304
Dublin, Ohio 43016
http://www.telemachuspress.com

ISBN: 978-1-948046-19-0 (eBook)
ISBN: 978-1-948046-20-6 (Paperback)

Library of Congress Location Number: 2018947846

Version 2018.07.12

For Family
Sandy, Mandy, Tammy, Kurt, Matt,
Dane, Thomas, Calla and Nathan

CONTENTS

PREFACE

Medicine and disease has preoccupied man from prehistoric times to the present. In all cultures and eras, populations have sought the means to preserve health and restore it when absent. The story of man's long circuitous journey to comprehend the human body, the origin of disease, and how to treat various infirmities is the subject of this book. An understanding of these older health practices can give insight and a better appreciation of modern medical miracles for both today's patients and those professionals that treat them.

The substance of this book is the compilation of multiple lectures that comprised a series of courses that I have given over the last ten years at the College of William and Mary's Christopher Wren Association. This is a group of volunteer instructors from a variety of backgrounds that provide life-long learning and personal enrichment classes to seniors in Williamsburg, Virginia.

The writing style and literary composition of this monograph is important to recognize. This is not a formal historic dissertation that follows the customary rules and guidelines of an academic research paper. Because the book's content is an amalgamation of many different oral presentations tailored for a lay audience, the typical annotations and footnotes used in a scientific manuscript that refer to a particular source, quote or author are not present.

The subject material is taken from the references listed at the conclusion. Some wording is paraphrased and borrowed without specific attribution from many of these resources. I claim no original work done except in my interpretation as an experienced physician of the information published by others. I am grateful to those many men and women authors who have researched and then written about different aspects of the history of medicine. Their efforts were an inspiration and collectively serve as the basis for this publication.

With the multiple millennia covered in this treatise, the discussion is sometimes blended into compressed chronological sequences in order to give coherence and a better understanding of the material presented. There are a number of subjects and individuals that were not examined. The people and issues that are proffered highlight and are representative of their era and importance in the overall development of medicine.

Finally, given the advantage of historic perspective, it is difficult to render past medical events in an unbiased manner; but I tried to do so and hope the reader will as well.

Jonathan L. Stolz, M.D.
Williamsburg, Virginia
May 2018

PART I
MEDICINE IN
ANCIENT CIVILIZATIONS

One of the hardest concepts to embrace is the large amount of time between the 21st century and early civilizations. While humans first appeared on earth over 200,000 years ago, it was not until between 15,000 B.C. and 7000 B.C. in the Stone Age that cave dwelling in France gave evidence of the prehistoric era. It took another 4000 years before Stonehenge was built—sometime between 3000 B.C. and 2000 B.C.

History truly began when the first writing in the form of the Sumerian cuneiform script appeared in Babylonia in 3200 B.C. Hieroglyphic writing during the time of the pyramids in Egypt emerged later in 3000 B.C. Ancient Egyptian society subsequently went through a number of transitional kingdoms for 2200 years until its end in about 800 B.C.

Ancient Greece lasted 700 years through three distinct periods—the Archaic, Classical, and Hellenistic—from 800 B.C. to 100 B.C. It ended with the fourth Macedonian War. Rome for 1000 years brought order to Europe. It began with the Roman Republic in the 6th century B.C. and lasted for 500 years until the Roman Empire started in the 1st century A.D. The fall of Rome occurred in 476 A.D. when Romulus, the last of the Roman emperors, was overthrown.

The time line of early India and China overlapped with Babylonia, Egypt, and Greece but was geographically removed resulting in a very different history of medicine in these cultures.

CHAPTER ONE
Prehistoric Era

P rimitive anthropoids of Homo sapiens have been on earth for several hundred thousand years. But as known today man is probably 40-50,000 years old. These early humans—our primitive ancestors—were hunter-gathers. They lived in caves sheltered from the rain and cold. Others resided in forests, on the savanna, or huts made of mud or foliage of which no trace remains. These prehistoric people subsisted on herds of wild animals and sought out plants to eat. Different primitive groups lived in widely scattered clusters of 50-100 people. They relied on each other for survival.

Those remains that have come to light are mainly from communal burial grounds. They reveal that early man was small in stature. At birth they often had congenital bony malformations similar to modern times: like hip dislocations and extra fingers and toes. Life was short lasting on average only 30 years.

While their lives were dangerous and precarious, their lifestyles may have helped them avoid some types of health problems that plagued people later on. Hunter-gathers did not stay in one place long enough to pollute water sources. This practice eliminated health risks associated with tainted water and diseases that were spread by insects living near contaminated rivers or lakes. There were no domesticated animals at that time. This excluded another potential source of ill health. In addition, since they lived in small groups and did not intermingle with others, contagious disorders were rare.

Because of their constant requirement to obtain food to survive, early man needed to maintain strong physical health and take care of injuries and illnesses. When a member of their group became ill or was hurt, the others had to devise healing methods.

Paleopathology

While there are no written records of the illnesses and treatments of these early prehistoric people, palaeopathologists have patched together a basic understanding of these early groups' health and lifestyles.

Paleopathology is the scientific study of the afflictions of prehistoric man using bones preserved in different types of soils and discovered in the course of archeological excavations. Because of its resistance to decay, bony tissue represents the majority of material available to palaeopathologists. By examining these skeletons using cutting edge medical techniques like DNA analysis, radiocarbon 14 dating, computed tomography scans, and electron microscopes, it is possible to identify some of the aliments that afflicted our ancestors. Evidence in fossilized bacteria, preserved in insect parts and parasite eggs, is also used to demonstrate diseases of that early era.

While skeletons cannot provide information about soft tissue maladies, it can often tell about a person's state of health. For example, a depletion of bone marrow iron indicates malnourishment. Additionally, the manifestation of rheumatoid conditions, tuberculosis, leprosy and syphilis were all seen in the preserved osseous material. Some of the remains revealed bone cancer and hereditary abnormalities. Mandibles showed poor dentition and signs of serious gum infection.

Beyond disease, hunting for big game, defending against predators and living in a harsh environment resulted in trauma. Palaeopathologists have found fractures of the long bones, spinal column, and pelvis. There were spearheads lodged in bones. Many of these injuries healed poorly even with primitive intervention. It left these trauma victims living out their short time on earth in a great deal of pain.

Prehistoric Medicine

Early prehistoric man needed to maintain good health and take care of injuries and illnesses in order to subsist. When poor health existed, cures were sought through trial and error.

But direct evidence about the medicine of early man is extremely meager.

Primitive people had no knowledge of germs and the role they played in causing illness. Sickness was attributed to curses or gods. They sought the best practical remedies they knew and often combined it with religious beliefs. Cave dwellings in France that date 17,000 years ago show art renderings of masked men wearing animal heads performing ritual dances—this is thought to be the earliest surviving images of medicine men.

It seems probable that prehistoric humans experimented with herbal medicines. They would have relied on a plant's appearance and color to treat a disorder. A bright yellow flower may have been used in an effort to rid a person of jaundice.

Although early mankind had little understanding of human anatomy, cave paintings depicted the location of the heart. In spite of this lack of morphological knowledge, well-healed fractures in prehistoric bones indicate that in some cases early populations became fairly adept in caring for broken bones.

In treating trauma primordial people had few surgical tools, no knowledge of the importance of cleanliness nor the means to prevent long term painful infections. There is evidence that they undertook amputations of injured arms and legs. Skeletal remains of dismembered limbs often reveal healing that indicates some people actually survived the excision procedure.

Other surgery did occur. The most commonly performed operation was trephination of the skull. It involved removing a small circular piece of the cranium to expose the tissue surrounding the brain. The motives for the surgery were likely for the relief of headaches, infections, convulsions or signs of insanity. Some palaeopathologists speculate that it may have been used as part of a magical or religious ritual to release evil spirits from the body. The risk of infection, brain swelling, or bleeding would have been high. A number may have managed to live after the procedure because only a small circle of bone was removed.

The primal methods of medicine that prehistoric populations employed to maintain health, treat illness and injuries likely had some positive outcomes on their daily lives and survival. Because writing did not exist, this inhibited any effective treatments from being passed down to future generations. It is clear nonetheless that maintaining good health was certainly as important to our primitive ancestors as it is for everyone today.

CHAPTER TWO
Mesopotamia

At the end of the last Ice Age 10,000 years ago, the world population was still mainly hunter-gathers, but other civilizations began to develop along riverbanks where there were well-irrigated lands.

One such area was between the Tigris and Euphrates Rivers. Here the ancient mighty empire of Mesopotamia developed in what is now part of present day Iraq. It flourished from the fourth millennium B.C. to the middle of the first millennium B.C. It finally collapsed from repeated assaults by different foreign conquerors—the last being by the rising power of the Persian Empire in 539 B.C. During its long existence the Mesopotamian population began written history and generated the oldest medical handbook known to man.

It was not an easy land to exist on. There were scorching winds, torrential rains, and unruly waters of the two large rivers. The people dwelled in sun-dried clay brick houses that often had to be rebuilt because of the wet climate. Those living in this tract flourished because of the great farming in the district. Overtime there were multiple different civilizations that inhabited the region. Mesopotamia became the name for the entire expanse of land extending from the mountains of Armenia to the Persian Gulf. In this vast area there was no political unity.

The first and most important civilization was the Sumerians in southern Mesopotamia. Their principle city was Ur on the Euphrates. The Sumerians overcame and merged with a northern empire called Akkad. This new amalgamation was subsequently replaced by yet another southern empire called Babylonia. Notwithstanding these different political vagaries in this large region, there is basically just one Mesopotamia civilization without a centralized government. This territory is often loosely called Babylonia, and the extended geographic unit is sometimes referred to as the Fertile Crescent.

Cuneiform Writing

A fair amount is known about Babylonian society. It is better documented over its history than any other ancient populations including Egypt. The emergence of writing in about 3500 B.C. was similar but independent of Egyptian pictorial writing. The ubiquitous clay in the region between the Tigris and Euphrates rivers was turned into tablets. Pictures and signs were edged into the soft but quick drying material. This ultimately led to a formalized script composed of many wedge-shaped lines that is called cuneiform. Of the over 30,000 surviving clay cuneiform tablets about 1000 concern medicine. From this group there is a collective portrait of Babylonian medicine in its final period of maturity during the first millennium B.C.

Among the many tablets recovered, there is an older one composed by an anonymous Sumerian physician who lived in the third millennium B.C. This ancient doctor collected and transcribed the most valuable medical prescriptions of the time for use by his colleagues. His three by four-inch tablet is the oldest medical handbook known to man. Most of the other health related tablets were also for prescriptions. A few described procedures like draining pus from the lungs. There was one that specified what fees could be charged for various physician services.

Mesopotamian Illnesses

The Babylonian populace was a static group that lived in cities and non-nomadic communities. This change in habitat from their roaming hunter-gather ancestors is what made the Mesopotamian population contract more serious diseases. For the first-time people lived in close proximity to each other for an extended period. This enabled airborne afflictions to spread from person to person with no intermediary carrier. The domestication of animals was the genesis of other major infections that were easily transmitted. Cattle likely started small pox. Pigs and ducks carried influenza, and measles was probably from dogs to cattle and then to humans.

The irrigation and farming in the region prompted an increase in infection carrying parasites. Disease laden insects lived near waterways that were polluted with animal and human waste. These bugs often entered the bloodstream of barefoot workers. This spread disorders like cholera and typhoid. Some people after exposure to various maladies built up resistance. Survivors then became a part of a healthier population. This improvement in the populace's over all wellbeing lasted only until the endemic diseases returned after the passage of time to infect those without inherent immunity.

One thousand years before the Christian era, Mesopotamian citizens lived in an environment dominated by gods and religion. Medicine became a part of this culture where numerous gods- and goddesses-controlled health and disease. The spirits were the companion of the gods and the guardians of the human body. Any infringement on religious rules that had been decreed provoked anger. Disease was endured as punishment for the various sins of impurity or uncleanliness.

One cuneiform tablet states: "Impurity has struck me. Judge my case, take a decision about me, eradicate the evil disease from my body, and destroy all evil in my flesh and muscles so that I may see the light."

Illness became entwined with faith. If a person committed a sin, the gods retracted their protection. Then the individual fell prey to the innumerable disease bearing devils that people believed swarmed around Mesopotamia. When this happened, specific names were given to the induced aliment depending upon the internal organ that was attacked.

Diagnosis and Therapy

Early Babylonian health care providers developed a process that led to a specific diagnosis of those who fell ill. They employed clinical observation of the sick person's symptoms and searched for which moral transgression caused the illness. The patient would be interrogated—similar to modern doctor obtaining a medical history—in order to discover what sin had been committed. If the malady was the work of a well-known devil, no further workup was needed. The prognosis was then deduced and the appropriate treatment applied.

In complicated cases divination (the process of seeking knowledge of the future by supernatural means) was utilized to discover the diagnosis and possible outcome of the illness. There were numerous divination methods available to the Babylonians including astrology, astronomy, and dreams.

Divination by studying the liver of a sacrificed animal is called hepatoscopy. It was one of the most popular methods used by Babylonian healers. Archeological excavations have found thousands of models of livers in baked clay, wood or bronze.

The liver was thought to be the seat of life, thought and feeling. The depth of the fissures, shape of the lobes and
consistency of the liver was evaluated by the healer to determine whether a patient's disorder was treatable.

Babylonian therapeutics were largely based on religious concepts. Confession and sacrifice with prayers were necessary to
appease the gods. Amulets, objects that had the power to
protect its owner from harm, were also common. Magic
spells and incantations were an important part of medical
treatments for Babylonians. They were incorporated with
medicinal recipes of the day. Sometimes highly poetic in nature, these conjurations were recited to drive out the evil
spirits of disease.

One medical historian postulates that the close relationship
between magic and medicine that occurred in Mesopotamia
has persisted into modern times but in a very different for-
mat. It is now called the "psychological dimension of therapy
in the healing arts." Scientific studies have shown that a patient often feels better on entering the doctor's treatment
room. In fact, the very sight of the physician's white coat and
stethoscope may in some instance make such an impression
on the patient that it becomes a part of the healing process.
This may be an updated variation of one aspect of ancient
healthcare that has in a small way entered into contemporary
medicine. On your next visit to your doctor's office check for
any amulets in the waiting room!

Medicine Beyond the Supernatural

Not all of Mesopotamian medicine was purely supernatural in approach. A number of the clay tablets contain descriptions of a disease, a diagnosis, and a prescription of drugs. Maladies of the liver, eyes, lungs and fevers were noted. They showed some knowledge of night blindness, otitis media (middle ear infection), kidney stones and strokes. The tablets promoted the virtue of cleanliness and that it was practiced widely. There were some sewage systems. Archaeologists have excavated 4000-year-old primitive toilets.

Medicinal herbs were prescribed. These included the poppy, the source of opium, as well as other plants like saffron, juniper, garlic and onions. The first mention of belladonna was in Babylonia. Its most important component, which was later recognized as atropine, proved to be a highly effective antispasmodic agent. Mesopotamians employed it to combat bladder spasms and coughing. While there were many other drugs and healing substances available in Mesopotamia, the question remains to what extent these materials were utilized for rational medical purposes.

Surgical techniques were few, but there were two procedures that the Mesopotamians appear to have pioneered. The first was the operation for cataracts. This delicate measure was called couching for cataracts in a process called "reclination." In this method a bronze needle is passed into the eye's pupil. The surgeon then presses the clouded lens downward into the lower part of the eyeball, out of the field of vision. The relocation of the opaque lens from visual path restored a degree of sight. It helped many patients.

The second surgical measure that Mesopotamian physicians were responsible for was catheterization of the bladder. This was performed in order remove obstacles of inflammation that interfered with the flow of urine. One cuneiform tablet reads "you can pass a bronze tube into the penis to stop obstruction and pass medicine."

Mesopotamian Doctors

So, who were the doctors during the Mesopotamian era? Historic evidence indicates that the art of healing was never in the province of a single profession during this time period. No one group had a monopoly. Diviners, priests, sacrificers, and doctors all treated the sick. Ancient Babylonian texts describe two main categories of providers who administered to the sick in different ways.

The first type was the *asu-physician* who was a layman entrepreneur. This group specialized in wound care that involved washing, bandaging and making herbal plasters. Their functions were also similar to an apothecary that prepared complicated recipes and drugs. The *asu* usually operated from a corner store, in the street or from a home. This kind of doctor could not enter the temple to practice his craft.

The second class of the healing profession was the *masmassu* or exorcist. These individuals had the social advantages of being priests. They were responsible for the magical prevention of attacks of demons and angry gods. This group determined which god or demon caused the illness and whether the illness was brought on by a sin of the patient. Some also acted as attending physicians and cared for patients on their sick bed in order to give a prognosis and predict the nature of the disease.

In the broadest terms, the Babylonian citizen would go to a *masmassu* or exorcist if one needed a ritual to undo a bad omen that caused the illness; but, if one had a nosebleed, then one would seek out the services of an *asu-physician*.

This highlights the independent approach medicine and magic took in healing and the complex relationship between the two. It was a dynamic process. It was not simply a matter of the primitive technology used by the *asu* to relieve symptoms. It was also through the *masmassu's* magic that would include the patient's participation in the healing process itself. For both classes of healers their concept of Babylonian medicine was primarily focused on how to alleviate symptoms of an already existing ailment. Neither group addressed the prevention of disease like through diet.

Additional information about those that practiced medicine is included within the Law Code of Hammurabi. Hammurabi was the ruler of Babylon from 1792-1750 B.C. who established this code of laws. It dictated the proper governance and rules for the citizenry as well as for the practice of medicine.

Part of this legal dictum stated that doctors were responsible for surgical errors and failures. The laws described punishment solely for lapses in the "use of the knife"—nonsurgical medical mistakes did not count. The punishment depended whether the patient was of high status or slave. For the former, the surgeon might have his hand cut off, but for the later the slave was to be replaced by the doctor. Fees for various services were also specified. This important judicial code was the first instance that the practice of medicine was regulated by the state.

The influence of medicine in the Babylonian culture is often overlooked and downplayed because they did not have any great scientific theories that have survived the test of multiple centuries. Still their doctors, treatises and rituals became widespread in those early days particularly in India and also served as the origin of Hebrew medicine. Most medical historians give some credit to Babylonian medical care. It was organized, pragmatic and served their population to the extent of the knowledge known at the time.

CHAPTER THREE
Ancient Egypt

Egyptian medicine evolved in a unique environment. The geography of Egypt was unlike any other country in the ancient world. It was an essential component in the historical and cultural events that allowed a sophisticated system of medicine, building, technology, writing and artistry to be developed.

As early as 10,000 B.C., settlements began to sprout along the Nile River in the area contained within the present political boundaries of modern day Egypt. The people found a plentiful amount of fish. In addition, the regular flooding of the river created fertile soil for farming that ensured a ready supply of food. The Nile offered the benefit of being navigable in both directions. By 3200 B.C. Egyptians established many trade opportunities.

The geographic location delivered protection from foreign intrusions. The broad expanse of desert beyond the Nile provided isolation from others. This made the region conducive to easy defense resulting in long-term equilibrium and a powerful sense of national identity for the population. Unlike in the Mesopotamian area where the different city-states never joined to become one political entity, Egypt's upper and lower kingdoms eventually united to become one country. The outcome was a 3000-year period of calm and governmental permanence. Egyptian medical achievements took place during this age of stability.

Hieroglyphics

As with other primordial cultures, some of the facts of ancient Egypt are based on information noted by writers from a different age. Because the Egyptians created a useful form of picture writing known as hieroglyphics, a contemporary lasting testament about Egyptian life and medical beliefs of the time is also available for study. The most important source of our knowledge of Egyptian medicine is predicated on papyri books written in hieroglyphics on material made from the stems of the papyrus plant. Much of these initial writings were collected in the library established in the city of Alexandria in 350 B.C. A fire occurred in 391 A.D. that destroyed many of these documents. Only a handful concerning medical practice have survived.

Those papyri available reflect the medical ideas of over 3000 years ago. The oldest known medical papyrus is the *Kahum*

Papyrus written in about 2000 B.C. It deals with gynecology and veterinary medicine. The two most important and informative are the *Edwin Smith Papyrus* (written in 1600 B.C.) and the *Ebers Papyrus* (written in 1550 B.C.).

The Smith document was found in 1862. It is a 110-page scroll that deals with surgery. It documents forty different battle wounds and how to treat them. It shows that the Egyptians had a relatively high level of medical knowledge along with empirical or experience-based treatments that had a magic component. The Ebers narrative is more of a medical text and explains a great deal about disease management. There are a number of other papyri but their content is largely restricted to prescriptions with a heavy dose of supernatural elements.

The papyri books that contained all this information were unique. The Babylonians etched cuneiform symbols into bulky clay tablets. But the inventive Egyptians found that by placing thin strips of the spongy tissue of the papyrus plant together then soaking, pressing, and drying them a sturdy surface on which to make notations resulted. The tough quality of the papyrus and the dry climate in Egypt helped preserve these documents.

Early explorers of the region were baffled with the strange picture writings. In 1799 French soldiers working in Egypt opened the door to their translation when they discovered the Rosetta stone. This stone featured two languages—Greek and Egyptian—and three different scripts: hieroglyphic, demotic, and Greek. It took 23 years to unravel the meaning of the hieroglyphics and unlock the Egyptian language. The Rosetta stone became the key to recast these and other documents.

The Doctors

Most people with pain, illness, or injury will turn to another for help. In any human society, however simple its organization some will emerge as possessing or appearing to hold superior curative skills. Inscriptions dating back to 2700 B.C. prove the existence of physicians, priests, and magicians who were all involved with healing especially in the court of the Egyptian rulers. Some even bore such poetic titles as "Shepherd of the Anus" and "guardian of the royal bowel movements." These names came about because the anus was considered the main seat of pathology, while the heart was the center of life.

The Smith Papyrus explains three hierarchical levels of professionals that practiced conventional medicine in ancient Egypt. The lowest class was the *swnw* (pronounced 'soonoo'), a word translated as physician. They were the doctors of the people. They were self-taught and learned through trial and error. At a higher status was the *wabw* (pronounced 'waboo')—a word translated meaning 'the pure.' They administered medicine to the more privileged in society. Religious rites were a large part of their duties.

The third tier was the *saw* (pronounced 'saoo') that means guardian. Their practice was limited to royalty and other elite

members of society. They usually gained their training within
the prestigious temple-palace schools. Others were from a
family of doctors where they underwent formal apprenticeships. They used the medical papyri like textbooks to
supplement their training.

While there was no system of medical licensure like today, the
Egyptian physician was required to practice according to very
rigid rules of treatment and in accordance with strict procedures. If a patient died, the doctor was usually absolved from
any wrongdoing. But if the physician deviated from the traditional practices and methods in any way, he was subjected to
the death penalty.

There was significant specialization found in the profession.
The practice of medicine was split into separate parts where
one doctor often treated only one disease. Some doctors
specialized in disorders of the eye, others of the head, teeth,
stomach and so on. This specialization suggests that the state
of the medical art in Egypt was quite advanced by ancient
standards.

It is difficult to pinpoint who practiced in the field of
gynecology and obstetrics. There is no evidence that the *swnw*
were involved in childbirth or even knew anything about the
normal conduct of labor. While it seems inconceivable that there
were no women involved or who had skills and experience at
assisting with childbirth, there is no documentation of their
presence.

Nevertheless, there were a few women who were a part of medical system. Females as a group had more status in ancient Egypt than in many other early civilizations. They could own property and had a number of legal rights. They served as nurses, wound dressers, and sometimes physicians. Women doctors tended to the health needs of other women as well as overseeing the use of health and beauty treatment involving creams and ointments. A tomb dated 2400 B.C. held the body of Lady Peseshet. She is thought to be the first female doctor.

Physicians had tests to determine whether a woman was pregnant. One involved placing oil on the woman's breasts and observing for the presence of dilated veins that occurs as an early sign of pregnancy. While surgery plays a major role in modern gynecology, there are no recommendations for surgical intervention in any of the papyri. Treatment was confined to oral meds and local applications mainly in the vagina.

The eye played a major role in Egyptian mythology. As a result, doctors became renowned for their skills and treatments of ocular disorders. While there is no direct archaeological evidence of eye disease itself, artificial eyes often replaced mummies' eyes, and there are many representations of blindness particularly of harpists. The Ebers papyrus notes prescriptions for medications to be applied externally to the eye. There is mention of the ingestion of liver to treat night blindness. It remains unknown whether Egyptian doctors actually recognized the direct connection between the eating of liver and its healing effect on decreased night vision. Today, it is understood that night blindness is caused by a deficiency of Vitamin A that is in high concentration in liver.

While there were many legendary doctors in ancient Egypt, the one that has stood the test of time is Imhotep. He was considered a God, but was in fact an historic figure that lived in about 2900 B.C. His name means "one who walks in peace." He is credited with accomplishments in many fields but most of all as a successful physician. He was one of the first medical men to stand out clearly from "the fog of antiquity" as one historian noted. He rose from the role of medical hero to become "God of Medicine." Outside of medicine, one of Imhotep's most lasting contributions was being the designer of the early Step Pyramid.

Anatomy and Physiology

While there is no surviving Egyptian works specifically on anatomy, insight into certain parts of the body in various medical papyri indicates that there must have been some study of the subject. There is no evidence that human dissection was undertaken in Egypt until after the Greeks arrived later and worked in Alexandria. Some limited sources of anatomical knowledge other than through dissections was available. Battle causalities and serious accidents were opportunities for the Egyptian doctor to learn about the fundamental human structures. In addition, embalmers had technical expertise, but it is uncertain whether they understood the details of the underlying anatomy or passed on what they observed to doctors.

Overall the anatomical and the physiological or functional knowledge of the internal organs were restricted. This lack of facts did not impede speculation about the elemental structure of the human body by the ancient medical practitioners. The Egyptian physicians believed that there were 44 different channels from the heart called *metu*. An appreciation and theorizing of the presence of these internal pipes is not surprising in a country where their very existence depended on irrigation canals. It was hypothesized that the beating heart distributed air, blood, mucus, nourishment, semen and excretions through these multiple channels to their respective organs like the lung, rectum, testicles and bladder. Egyptian doctors postulated that obstruction of any of these internal canals in which substances traveled would cause disease. In order to cure the sickness, the early providers believed it was essential to eliminate the blockage.

Closure of the intestinal channel was thought to be particularly dangerous. Egyptians were obsessed with bowel movements and the prevention of the lower channels or *metu* from becoming clogged with excretions. A solution to the problem was found when a doctor observed a large bird, the ibis, standing in the water of the Nile, introducing its long beak into its rectum. It was assumed that the bird was washing out its insides. This was the origin of the clyster or enema. The Egyptians invented syringes and used this procedure liberally in

their treatments of multiple different disorders and not just for
bowel blockage.

Diseases

Most early civilizations had their own unique explanations for
the causes of different illnesses. Like those in Mesopotamia,
the Egyptians believed that everyone was born healthy, but
evil spirits sometimes entered the body to cause disease.
Religion and magic, in which there was little distinction, were
regarded as major controlling factors in the events of daily
life. If the cause of sickness was thought to be supernatural,
certain deities were invoked for cures. So, some patients
sought the help of the gods in preference to doctors. The
motive for choosing a supernatural cure over the then
available medical therapy included cost, risk, uncertainty of
outcome, and the sheer physical pain of ancient treatments.

The reason for the sickness would also have been an important factor in the patient's decision to choose between
conventional medical treatment and religious magic. The
cause and outcome of trauma was usually predictable making
traditional medicine the obvious choice. Magic or religious
incantations were rarely used for injuries. In contrast, the
etiology of internal disorders or "hidden disease" was usually
unknown and the outcome was uncertain leading the patient
to use supernatural cures.

It is often assumed that the ancient Egyptians must have suffered from many of the same range of disorders that effect that country today. Evidence from human remains, representations in art, and diseases in the medical papyri suggest that this may be true for degenerative disorder like arthritis and arteriosclerosis. The incidence of cancer was less than today most likely secondary to the shortened life span of the populace. Infectious disorders, while certainly present for those ancient people, were of a different character than in modern times. Other medical conditions are difficult to assess from the information available on the various papyri. Certainly digestive, pulmonary, and urogenital abnormalities were seen within the population because treatments for these conditions were described.

Parasitic infections were surely present then like it is today. It is difficult to prove this or the appearance of any other bacterial or viral afflictions using mummies or skeletons. There are few clues in the papyri nor in the illustrations in the tombs depicting typical inflammatory symptoms. It does seem likely that tuberculosis was present. The typical features of one aspect of that disorder that manifests as a humpback in an individual when the malady involves the spine is depicted in art. This variation of tuberculosis is called Pott's disease. It is also reasonable to assume that there were abscesses of various kinds because of the environment in which the people lived. Lastly, deformities like dwarfism, clubfoot and hydrocephalus persisted in mummified and skeletal remains.

Pharmacopoeia

The ancient Egyptian pharmacopoeia was weak by modern standards. The basis of treatment was mainly empirical rather than rational. In most cases cures were aimed at the relief of symptoms rather than eradication of the cause of the disease. To a large extent this was inevitable since the nature of most internal complaints was not known at the time.

The therapeutic options included: TLC, magic, invocation of the deities, improvement of diet, massage, and surgery. By far the commonest form of treatment recommended in the medical papyri was the use of an assortment of different drugs. These medications would be drawn from a wide range of animal, mineral and vegetable substances dispensed by volume, not by weight. They would be given to the patient in a variety of ways including oral, rectal, vaginal, external application, and fumigation. The ancient Egyptians were renowned for their skill in administering drugs. The nearly 900 different prescriptions noted in the papyri were made from over 500 different substances that were combined together for specific results. The main liquids used as the vector for these medical mixtures were water, milk, wine, and the urine of an elephant cow.

Some substances like the drug Senna and castor oil were effective to purge the system. Tannin was used for burn therapy. More bizarre ingredients by today's standards included animal dung and lizard blood. One treatment for curing baldness had its curious origin from the hedgehog because the animal's body is covered in spiny hairs. This measure was used in combination with exorcism that had the bald man being treated recite: "Oh, you the luminous one who does not stir from your course, you who fight against transgression, beware of him who has made himself master of the top of his head."

Another interesting creation was a vaginal contraceptive solution made of crocodile oil, honey and natron that was a substance used in embalming. When these ingredients were dissolved together, an acidic solution resulted. It would have been an effective spermicide.

Workers building pyramids received medical care. They were given enormous amounts of garlic, radish and onions. Today, it is known that antibacterial and anti-fungal compounds can be extracted from these vegetables. Perhaps as a result, internal disease in these laborers was reduced. Challenging work conditions involving extensive exposure to the sun and blowing sand produced a variety of eye ailments; and broken bones were a frequent problem for those laboring with massive stones.

Some historians have speculated that the workforce may have benefited from a form of early health insurance. In this instance, one papyri mentions that incapacitated injured workers received job exemptions or assignment to light duties and were even given a pension. Perhaps modern disability insurance had its roots along the ancient Nile or more likely this is a case of modern thinking and concepts applied to an ancient practice.

Overall most of the drugs available were purely sympathetic remedies. In some cases, there were limited therapeutic effects; but this was uncommon and usually unexpected. The best example of an unforeseen result was the Egyptian physician's use of solutions from the leaves and bark of the willow tree for "fever." Analysis has shown that concoction has the same properties as aspirin.

Surgery

There is no convincing evidence that the specialty of surgery was practiced separately from that of general medicine in ancient Egypt. The management of injuries was excellent and well documented. There is little knowledge about surgery being used to relieve conditions that did not result from traumatic damage to the body. In a report of the examination of 30,000 mummies, not a single surgical scar is noted except where there may have been trauma. Further in contrast to the Graeco-Roman period, no one has found surgical instruments from the Egyptian era.

For injuries, the Edwin Smith papyrus is a text for the management of wounds. It describes in detail 48 illustrative cases starting at the top of the head and proceeding downward in an orderly fashion. Lacerations, fractures, and burns were all treated with a remarkable degree of success in a logical and impressive approach to care. There is a large discussion on how to effectively treat scorpion stings and both snake and crocodile bites. Space does not permit detailed discussion of all their groundbreaking technics, but suffice to say Egyptians were likely the first trauma surgeons.

Mummification

When one died in ancient Egypt, the preservation of the resulting dead body was paramount. There was a deep belief in life after death and a desire to be certain that the deceased would have access to his or her body in the hereafter. Early on Egyptians buried their dead in shallow graves in the desert. The hot, dry air proved to be a perfect natural mummification process. When burial inside tombs and pyramids developed, the bodies rapidly decomposed; and the search began for ways to preserve the corpse so that it could be reused in the afterlife.

Experimentation with naturally occurring compounds resulted in a successful mummification method using natron. This chemical, that was present in dried lakebeds, proved very effective in drying out a dead body and preventing decomposition. Natron is composed of sodium carbonate, sodium bicarbonate, sodium chloride and sodium sulphate. There were three embalming methods that varied in thoroughness and cost. These services were available in a range of prices—not unlike today's funeral parlors.

The top of the line method involved the removal of the internal organs. A hook was used to remove the brain through the nose. The internal organs, except for the heart that was considered the "seat of the mind," were removed through

small incisions and placed in jars. The empty cavity was then
washed with wine, myrrh, and spices to provide fragrance.
The corpse then was treated with natron and wrapped in
linen that was covered with gum to create a glue-like casing.
After 70 days the mummified body was returned to the family
to be placed in the burial chamber. The jars filled with the
internal organs were buried with the mummified remains.

A less costly method of mummification simplified the number of steps by filling the cadaver's abdominal cavity with
cedar oil injected through the rectum. This caused the inner
organs to liquefy. The body was then covered with natron for
70 days. At the end of this period, the oil was permitted to
drain out of the body, leaving only skin and bones. The final
and cheapest method was to wash the corpse in a salt solution, embalm with natron for 70 days, and return to the
family for burial. Mummification was used until the seventh
century A.D. It survived into early Christianity but ceased
with the beginning of Islamism.

Studying the mummies gives valuable knowledge to modern
palaeopathologists from which they learned about the early
Egyptians. The mummies also serve as mute witnesses to the
diseases of those living along the Nile and the medical practices of that great ancient civilization.

CHAPTER FOUR
Ancient India

There are three distinct timelines in ancient India. The first is referred to as the Harappa civilization. It existed as early as the third millennium B.C. Like in Egypt and Mesopotamia, these early settlements began along major rivers. They developed a remarkable culture that was distinctly independent in nature. It disappeared in approximately in 1500 B.C.

After that early settlement, there were two other epochs. The first lasted until about 800 B.C. and is called the Vedic period. Information from this era is derived primarily from four holy Sanskrit books called the *Vedas*. The final age, from about 800 B.C. into the first millennium A.D., is called Brahmanic because it was based on that caste of the Hindu priests. It was during this time that Ayurvedic medicine blossomed.

India's medical progress after the early period was very much entwined with the religious developments of the country. It was initially influenced by the development of the Hindu, Brahmanism and then by the unfolding of Buddhism.

Harappa Era Medical Practices

From 2300 to 1500 B.C. the advanced ancient civilization of Harappa was located in the Indus River valley. This area is in present-day Pakistan and western India. Their culture was highly stable and not preoccupied with warfare. It was known to have a well-developed agricultural community that provided food for the populace.

For the early Indians water was the source of all life and considered a powerful purifying agent. Because of this, in their cities the Harappa people developed community facilities for sanitation that augmented their belief in the importance of hygiene. Rubbish was disposed of in outlying areas, and many homes had inside water pipes including a bath and toilet. This was associated with an extensive drainage system that carried the wastewater in covered sewers away from the residential areas. All of this helped reduce exposure of the community to some illnesses. This sophisticated level of very early public sanitation was equal to what Roman authorities would eventually achieve 1000 years later.

There is scant information on the medical practices of this early civilization. Much of what is reported in the historic literature is reasonable speculation based on archeological findings. There is no holy grail like the cuneiform tablets of Mesopotamia or the papyri from Egypt to provide solid facts on Harappa medicine.

It seems probable, however, that the healing system used by the Harappa culture was similar to the other great civilizations of the time. This suggested that medicine was inextricably connected to religious beliefs and based on magic. Certain aspects of healing involved séances conducted by medicine men called shamans. During these rituals, it is reasonable to assume, plants along with powerful incantations and ritualistic dances were used to exorcise diseases that resulted from demonic possession. Unique to the Harappa population was the employment of water related activities like bathing and ablutions because they believed in its purifying properties.

One historian reports that there is evidence of smallpox in the Harappa region as early as 3000 B.C. Healers were thought to use a scarification process to inoculate people against the disease. India was one among several other ancient places where this activity was utilized long before it was performed in western societies in the 18th century.

Perhaps, the most enduring health tradition of Harappa medicine is Yoga. Its earliest known origin dates to the Harappa civilization where yoga positions are depicted on some of the artifacts found in the areas where they lived. After the Harappa population disappeared from the Indus River valley, the practice of Yoga appealed to the Hindu people that eventually inhabited that region. Then, as aspects of the yoga evolved, it became integrated into the Hindus' good health habits centered on a lifestyle of control and self-discipline. The term yoga is from the Sanskrit word for yoke. This generally is interpreted to mean, "to yoke or harness the mind" since concentration is a key element in yoga.

The Harappa civilization slowly declined as changes in the course and level of the rivers where they lived coupled with erratic climate patterns to reduce water supplies. This led to a decrease in crop production and eventually to economic collapse and social decay within the population.

Vedic Age Medical Practices

As the Harappa ethos disintegrated, the Indo-Europeans migrated into the Indus River valley where they destroyed the remaining members of the Harappa civilization. This new group called themselves Aryans. They started the Vedic age that lasted from about 1500 to 800 B.C. The Aryan culture became the early followers of the Hindu creed. They composed the oldest scriptures of that religion called the *Vedas*. Although it was a devotional work in nature, the five chronological components of the *Vedas* gave a comprehensive picture of the history of Vedic India.

At its inception, this new civilization was a pastoral society located in the northwest part of India. It spread elsewhere in semi-nomadic tribes led by warrior chieftains. Eventually they became more sedentary and agricultural in nature. In this setting a hierarchy of four different social classes developed that was the precursor of India's distinctive caste system.

Much of the Vedic medical practices were documented in the various *Vedas*. In these writings an entire religious and intellectual philosophy of life is revealed. This thinking incorporates one of the distinguishing features of ancient Indian medicine: that is, the impossibility of distinguishing the difference between the soul and the body and also the dissociation of the visible from the invisible because both are energized by the same life force. These old texts furnish information on the treatment of wounds, snakebites, illnesses and surgery. While much of the material was written long after the practices that were described occurred, the data was based on the oral tradition of Vedic medicine that transferred information from teacher to student and one generation to another.

Like so many other early societies the Vedic's basic belief was that actions determine destiny and that sin was the cause of disease. As a result, the importance of leading a good and moral life was the focus of early Vedic medicine. The understanding of disease was twofold. First, maladies inside the body were caused by invisible demons. Second, they believed afflictions on the outside, like wounds, were derived from external visible causes. This concept was a step forward for medicine.

In spite of its religious trappings, it is possible to recognize in the various Vedas books sufficient symptoms of different diseases to determine what illnesses were suffered by the Aryans. Not unexpectedly fever-type maladies predominated. Malaria flourished in the tropical dampness of the river valleys where the ancient populations lived. The region was a breeding ground for other contagions like forms of plague and cholera.

There are symptomatic descriptions that are detailed enough to recognize gout, epileptic fits, abscesses, jaundice and many other disorders. Dropsy—the old fashioned term for edema—was described in three words, "water in belly." One medical historian observes with some surprise that the Vedic doctors had been aware of the connection between dropsy and cardiac complaints. Edematous swelling of the lower extremities was, in some reports, connected with pains in the chest by the ancient Indian physicians.

In Vedic medicine diagnosing the cause of one's affliction was conducted with a strong emphasis on observation that involved isolating and identifying dominant and recurring symptoms. This process was attempted even though the Vedic healers possessed only a superficial understanding of human anatomy. While anatomical information may have been obtained during sacrificial functions, this was exclusively religious in nature and part of the priestly domain. Any potential information for medical use was not shared with Vedic healers.

Good health meant careful hygiene by regular bathing, good diet, and exercise. Even in this early period the Hindus showed a predilection for the purifying aspects of water as a cornerstone of good health. They specified the amount of water to be consumed before and after a meal. They also believed in cleansing their digestive tract with purgatives and water enemas.

The treatment methods for the internal demons that produced illnesses in Vedic medicine heavily relied on religious beliefs and specialized healing rites. During healing séances, the evil spirits were driven away with various chants while fragrant plants were burned to ward off malevolent forces. During the process the *Vedas* sometimes used intoxicating drugs from plants to soothe any pain. The modern equivalent is unknown. Therapy for external afflictions and poisoning took a more rational and empiric approach involving primitive forms of surgery including amputations and the use of artificial limbs.

While the Vedic age was instrumental in the birth of the Indian civilization, its early role in the development of medicine was greatly eclipsed by the advancement that doctors made in the Brahmanic era that followed.

Brahmanic Period

The Brahmanic period lasted about 1000 years beginning in 800 B.C. The greatest medical advances and ideas of the classic Ayurvedic Indian medicine were developed between 700 and 200 B.C. The word "Ayurveda" is a derived from *ayur* that means life and *Veda* that means knowledge. While

Ayurvedic medicine grew from Vedic traditions that preceded it, it was more empirical in nature and dropped many of the magical components used previously.

The teachings of Ayurvedic medicine were compiled in two classics sources written in Sanskrit. They were based on older works with numerous additions made over time. The first, *Caraka Samhita*, was a collection of writings of the healer Caraka who lived sometime between 1000 and 800 B.C. This work is thought to have been the first treatise of Indian medicine. It was so influential that it became a standard textbook for almost 2000 years.

The *Susruta Samhita* followed the *Caraka Samhita* . The former was composed in the fifth century B.C.—about the same time as the height of Greek medicine. Authored by a well-respected medical practitioner Susruta, the manuscript was a compilation of various advanced types of surgical interventions being performed in ancient India.

Ayurvedic Doctors and Clinical Approach

The Indian doctors in this time period belonged to a third caste. This was lower than the priest and warrior castes. They were called *Vaidya* that means "he who knows." Their medical education was of high caliber. It did not take place in priest schools but through apprenticeships with gurus. There was a balance between theory and practice as well as between medical and surgical teachings.

Introduction into the profession was marked by a solemn ceremony in which a pledge, similar to the Hippocratic oath, was taken. This is one of the many parallels between Hindu and Greek medicine. These early Ayurvedic practitioners had to vow to be celibate, speak the truth, eat vegetarian, never carry weapons, obey the master, and pledge devotion to their patients, including not having sex with them. They were also forbidden to help enemies of the King and could not treat women, unless their husbands or guardians accompanied them.

Dissection of a corpse was not permissible. The Vaidya's knowledge of the body was therefore very poor. Nevertheless, some anatomy was learned from an unusual activity according to records in the *Susruta Samhita.* Through an elaborate process, a cadaver in a cage was placed into a river. After a week it was removed and the outer skin was peeled away to study the ligaments, bones and some of the internal organs. While some superficial knowledge was obtained from this practice, it did not bring about any significant improvement in anatomical awareness.

While the supernatural elements in Brahmanic medicine were considerable, there was also a surprisingly speculative semi-scientific role at play. It was theorized that the body was composed of seven living substances: blood, chyle, flesh, fat, bone, marrow, and sperm. Good health depended on their equilibrium. Like their Greeks counterparts, Brahmanic healers believed symptoms were caused by too much or too little of these body humors and body elements. Under this hypothesis the therapeutic art of the Indian doctor consisted solely of re-establishing the harmony that reigned prior to the illness.

Diagnostics were highly developed. To arrive at what was wrong with the patient, the Indian healer initially listened closely to the person's description of the illness and the type of pain if present. This was followed by a thorough inspection of the patient's general appearance. Touch, palpation and auscultation of the body's internal noises were an important part of the process. After this the patient's blood, body fluids and excretions were examined. The last part of the evaluation was determining the prognosis of the illness identified. This was very important since Indian healers in general did not treat incurables.

The clinical knowledge of the classic Indian physician was very extensive. He was aware of blood spitting in pulmonary consumption or tuberculosis and the abdominal swelling or ascites of liver disease. They knew about malaria and perhaps even the role of mosquitoes in the transmission of the disease. They recognized that plague like epidemics were preceded by rats dying in great quantities. Through all this, these early Indian doctors were obsessed with the classification of different illnesses. For example, they enumerate 66 diseases of the oral cavity and 5 of the earlobe.

While all forms of therapy included prayers and spells, there was an extensive non-religious approach to treatments as well. The center of the doctors' therapeutics was the correct diet.

The properties of various foods were closely examined. Indian practitioners early on recognized that a salt free diet was indicated in treating hydropsy where there is an abnormal amount of fluid or edema in the abdomen or lower extremities. They also liberally used cathartics, venesection and leeches.

Ancient Indian medicine developed a very extensive pharmacopoeia. Herbs as well as drugs of mineral, animal, or plant origin were used. The spirit in which the drugs were applied was important. The *Susruta* states that the remedy should not be stronger than the patient or the disease. The Indian Ayurvedic medicine was notable for the philosophy that slow steady cures worked. The herbal remedies were the kind that assimilated gradually within the body. This resulted in fewer side effects. The considerable fame of these Indian drugs led to their adoption by many other civilizations.

Surgery in Early India

The most brilliant treatment area of the early Indian physicians was undoubtedly their expertise in surgery. Their surgical prowess showed innovation as well as comprehension of what needed to be accomplished to correct an abnormality. Surgeons excised tumors, drained abscesses, extracted foreign bodies, repaired anal fistulas, and splinted broken bones. Like in other civilizations, they performed cataract surgery with a high degree of success. In the *Susruta* , there is the initial historic account of a tourniquet's use in clinical practice.

Archeological discoveries document over 100 different surgical instruments available. Cauterization with hot irons and caustic salves were employed. A great number of bandaging materials were known. Even relatively successful antiseptic care was utilized well before it was accepted as routine in the 19th century. This application occurred because of India's cultural emphasis on sanitation and cleanliness. This same hygienic philosophy extended to the operating table where infections were reduced. There is reference to drinking wine before surgery and one text refers to burning hemp near the patient before a procedure began—maybe a primitive attempt at anesthesia to reduce pain. To obtain proficiency in surgical skills, the *Susruta* recommended various experimental methods to learn new techniques. To enhance their dexterity physicians practiced incisions and excisions on vegetables and leather bags filled with mud

There were several areas in which the Indian surgeons truly excelled and accomplished successes that were unheard of until much later in time. The bladder stone was one of the most painful afflictions of humanity and for ages was regarded as fatal. The ancient India doctors perfected one of the first surgeries to successfully remove a bladder stone. This procedure did not become commonly available to Europeans for another 2000 years. Indian surgeons learned to repair torn intestines using an innovative "stitching" method that employed large black Bengali ants to bite the borders of both sides of the wound together. Once the ant had latched onto the edges of the damaged segment, the head of the ant was removed without disturbing the mandible that remained behind and kept the wound closed.

The acme of Indian surgeons' innovations was reached in their mastery of cosmetic surgical operations on the ear and nose. Doctors perfected the basic methods of this delicate surgery. They repaired defects of ear lobes torn because of Indian women's custom of wearing heavy earrings. They also instigated nasal remodeling or rhinoplasty. Many of these operations were performed on women who had part of their noses removed as a punishment for disloyalty. There is little doubt that the plastic surgery that flourished later in Europe was a direct descendant of these classic Indian techniques.

Indian medicine and surgery exceeded many of the medical practices of other early civilizations. The Indians were more scientific and often better technicians, but they never achieved a complete separation and division of medicine from religion. That severance would not come until the era of the early Greek civilization that would incorporate many of the Indian treatments into their own unique style of medicine. Perhaps the most enduring legacy of the ancient Indian healers is that many of the traditions of Ayurvedic medicine is still used today.

CHAPTER FIVE
Ancient China

A kin to the health developments in India, the Chinese medical explorations took place in isolation and also proved to be quite effective. Traditional Chinese healthcare is one of the world's oldest forms of medicine. Until 1972 when President Nixon went to China, most American practitioners knew little about it. His trip and the opening of that region of the world resulted in a greater appreciation of modern and ancient Chinese medical practices.

Early Chinese civilizations were separated from the rest of the world. Neolithic settlement in 6000 B.C. began along the rivers. Their history continued into the second millennium B.C. when the ancient emperors and dynasties dominated and made advances equal to that of the Mesopotamians. The earliest Chinese writing known to exist dated from 1200 B.C. and was inscribed on bones. Osseous pieces were carved with characters in a script not fundamentally different from the

Chinese writing of later periods. Information on the living conditions, weather, illnesses and medical treatments were recorded.

Over many centuries there were a number of disorderly shifts in China's political powers. Through all the chaos the Chinese people made astounding discoveries that often-outpaced western cultures of the same time period. As early as 100 B.C. the Chinese created paper for writing. An effective ship's rudder was invented 1200 years before any similar object was seen in the west. The Chinese also developed the first magnetic compass and a system of cartography based on grids.

Early Chinese Medical Theory

Chinese medical literature was very extensive. The genesis of medicine is credited to three legendary emperors. Fu-Xi invented the fundamental philosophy of *yang* and *yin*. Shen-Nong in about 2700 B.C. invented acupuncture, and Huang-ti in 2600 B.C. is the author of *Nei Ching*, the classic book on internal diseases.

Beginning in about 1750 B.C. the philosophy of Chinese medicine rested on the classification and the inter-relationships of the body to the universe. The Chinese believed that the whole cosmos is divided into two methodological classifications, the yang and yin. These were two inseparable forces. Yang was associated with the heavens. It was bright, dry, and masculine. Yin was characterized as dark, moist and feminine and associated with the earth. They were not in opposition to each other like bad and good, but rather were complimentary. The presence of both forces was considered necessary for a proper balance to occur. Pure yang and pure yin could not exist. Put another way, there is always yin in yang and yang in yin.

Within the body yang has to do with action and transformation. The intestines, gallbladder and stomach were yang organs. Yin on the other hand relates to circulation, nourishment and growth. It is associated with the heart, liver, lungs, kidneys and spleen. Interestingly, the brain is not mentioned because like other ancient people the Chinese did not understand its importance.

The Chinese were fascinated with numbers. They classified almost everything in an orderly fashion divided up in groups of 5. There were 5 basic elements (wood, fire, earth, metal, and water), 5 directions, 5 seasons, 5 sounds, 5 body parts, 5 planets etc. These groups worked in concert creating a dynamic system of inter-relationships: like water puts out fire and fire melts metal.

Early Chinese medicine considered that anatomy had little value. No anatomical research was practiced because ancestor worship forbids mutilation of the body of a dead person. As a result, with no anatomical knowledge, the Chinese only were able to envision the physiological functions of the human body as imaginary and fanciful.

This led to the disease theory of Chinese medicine that was dominated by a very formal and elaborate system of natural philosophy. For the Chinese doctor all infirmities were regarded from the perspective of the *Yin-Yang* pair and their intercommunication and connection with the various 5 components. When these multiple elements were in disharmony, there was disease. The resulting malady was thought to connect and interfere with the planets, seasons, colors and sounds that corresponded with specific organs in the body.

How Medicine Was Practiced

There were distinct classes of those who practiced medicine. The scholar-physicians mastered the subject by studying written material followed by an apprenticeship with an individual from a family of healers. Other practitioners were more specialized and focused on such things as food, simple disease, elementary surgery and animals. There was a category of untrained lay healers. They were often deemed charlatans. Women serving as midwives or wet nurses were considered to be of a lower status. All levels of Chinese physicians did not like caring for those who had a poor prognosis or were considered incurable because of potential punishments if their treatments failed.

Even from the earliest times, specific details of a disease were not the primary focus of Chinese medicine. As implied in their complex interconnected origin of disease philosophy, doctors considered the patient as a whole. They shunned treating a disease in isolation and worked toward an overall total body cure that eventually would rid the person of their illness.

As a consequence, the doctor's diagnostic process of the ill patient was not a simple procedure. The physician initially used his power of observation. A patient's behavioral history and review of the current symptoms was obtained. This included an understanding of the pain suffered, appetite, digestive disturbances and the quality of sleep. All bodily functions were also studied.

One of the most pivotal diagnostic procedures used by Chinese physicians was evaluating the pulse. It was so essential in pinpointing illness that the early Chinese people were said to describe a visit to the doctor as going to "have their pulse checked." Understanding one's pulse was considered a method to assess the circulation or *chi*. The Chinese believed this life force moved through a series of channels beneath the skin. The pulse was important because it reflected the link between the cosmos and man.

The pulse examination was a very elaborate process that took up to three hours to complete. The doctor examined three different points along the radial artery of the arm. He was expected to recognize 50 different types of pulses and more than 200 variations of rhythm and regularity.

After the completion of the initial observational process and pulse evaluation, the doctor was ready for the physical examination. The scope of this part of the workup was severely restricted. While the physician would take the pulse and look at the patient's eyes and tongue, other physical contact was kept to a minimum. To help in the assessment patients were encouraged to indicate the location of their symptoms by pointing to an anatomically correct ivory figure that the doctor supplied. Female patients were usually concealed behind a screen and unseen by the doctor. A servant or the woman's husband served as an intermediary and noted the location of the pain or discomfort of the woman on an ivory figure as well.

This elaborate assessment system lead to many different recurrent diseases being observed. While not labeled with modern terminology, infectious disorders like dysentery, measles and cholera were well described. Similar to some other ancient civilizations, the Chinese medical system suffered from containing too much detail and over elaboration. For example, no less than 37 different shades of the tongue were described and 42 distinct forms of small pox were recognized.

When the Chinese medical professional's appraisal was concluded, the findings resulted in a prognosis of the patient's problem. This then led to the physician's recommendation for a treatment. In almost all cases, the first mode of ministration would be a strong suggestion for the patient to maintain a code of good health by following the basic precepts of hygiene and a healthy dietary regime. Next, the doctor would choose medicine not for its efficacy, but on the basis of a standard code adopted in accordance with universal harmonics drawn from the 5 senses, 5 colors, and 5 smells.

The objective was to remedy the discords between yin and yang and the different 5 elements associated with the malady. By using the right drugs, healers hoped that the person could fight off illness on their own. There were no quick cures here!

Medicinal Cures

The Chinese *Materia Medica* (medicinal cures) or pharmacopeia was highly developed drawing on the vast resources available in China's broad geography. The majority of the 1800 available drugs were unlikely to have worked any better than as a placebo. But there were those that may have been effective.

In the second millennium B.C., a physician could recommend a drug (*ephedra sinica*) derived from the joint-fir for coughs and lung ailments. A derivative of it is still used today. Iron was given for anemia. Ginseng was supplied to invigorate a patient. Modern alternative medical resources provide it for fatigue and as a diuretic. The Chinese were apparently the first to use mercury in the treatment of ulcers of various etiologies—2000 years before it was prescribed for the treatment of syphilis in Europe. The "Flower of Heaven" as the Chinese called smallpox was observed not to recur in those that survived one bout of the contagion. To achieve this protection in others court physicians artificially induced mild cases in healthy people from pulverized scabs of infected patients. It was administered as a powder via the nostrils. This may be first instance of a documented smallpox variolation—a true innovation. The process was seen elsewhere outside China but was not in common use for centuries.

Tea drinking became a popular Chinese practice and assumed a reputation of promoting good health. Most likely it was not the tea but rather the boiled water that was helpful because it reduced the incidence of water borne pathogens.

Surgery and Acupuncture

With their culture's deep-rooted aversion to the shedding of blood combined with a belief that mutilation continues in the afterlife, surgery never developed among the Chinese. This was in spite of their considerable knowledge of anesthetics and their otherwise extensive adoption of Hindu medical practices. The most common act of surgery that the ancient Chinese performed was not for the treatment of disease. Because the royal court employees were eunuchs, surgeons were called upon to perform castrations. They became quite accomplished in this procedure.

Chinese physiotherapy was highly proficient. It used methods like massage and gymnastics. The Chinese today are well known for two other healing methods that date back to the ancient era: acupuncture and moxibustion. There is no credible date for the original use of either, but for acupuncture it is referenced in second century A.D. Chinese medical texts.

Acupuncture's healing method stems from the Chinese theory that each body organ is linked to a specific area of the skin that can be readily accessed. There are needles of different sizes and for specific ailments. They are inserted at a particular point depending upon the nature of disease and the organ affected. The Chinese believe that the procedure restores the body's balance of energy or *chi* and returns the person back to normal health. Modern neurophysiological research has found evidence that endorphins (chemicals that interact with the brain to reduce pain) are produced following stimuli from various areas of the skin. This may be responsible for the success some receive from acupuncture.

An interesting sideline story occurred when Henry Kissinger made an advance visit to China before President Nixon's 1972 historic trip. James Reston, a reporter for the New York Times, was part of the press contingent accompanying Kissinger. Reston became seriously ill and required an emergency appendectomy. The surgery was successfully performed at a Chinese hospital using acupuncture instead of anesthesia. To the amazement of the American visitors (and also happily for Mr. Reston) it proved quite effective. It helped initiate a cultural exchange of doctors between the US and China.

Moxibustion was a counterirritant process that was first used in 500 B.C. Heat or burning embers from a combustible plant is applied to the skin. This causes a cutaneous blister that is thought to stimulate circulation and restore balance to the body. It effectiveness is unknown, but it has been used for centuries and by some today.

The Chinese had many unique features in their early system of healthcare. For many historians their collective results to cure the sick were not much different than many other ancient civilizations. Like with physicians from different cultures with a similar state of knowledge ancient Chinese doctors depended upon the recuperative powers of the body and the self-limiting nature of many acute illnesses to produce results.

Until recently China functioned in isolation with its early medical history virtually unknown to the world. In spite of this its unique legacy has persisted. Some of the beliefs that were put forth by the ancient Chinese have survived and are practiced in parts of the world today as holistic or integrative medicine.

CHAPTER SIX
Ancient Greek Medicine

Ancient Greek medicine is incomparably closer to modern medicine than any other historic form of healthcare. Many aspects of contemporary medical practices would
not have developed without the precedents established by
that age-old civilization. In a period of about 1200 years
Greek physicians generated a ferment of continuous change.

Their legacy and ground-breaking premise was that disease
should no longer be regarded as a supernatural phenomenon.
Under the Greeks, those practicing medicine realized that
healing religious customs had limitations. Ancient Greek
medical care became rooted in an approach that embraced a
rational, naturalistic, and scientific point of view. Their progress, however, was not linear. There were times when slippage
occurred and a state of flux between science and religion was
evident. In the end, those lucid ideas that came to the forefront

had a far-reaching effect in medicine throughout the western world.

The history of the Greek civilization was not one of a unified ethos but a collection of peoples scattered around the Mediterranean region. Most historians date ancient Greece from the first Olympic games and divide it into four chronological divisions. The first is the Archaic period from 750-500 B.C. This time span is steeped in the mythological legends of Apollo and Asclepius, the god of healing. Next is the Classical period from 500-325 B.C. This is the era when the Parthenon was built and a series of political changes and power shifts occurred. Two major wars shaped this age: the Persian Wars and the Peloponnesian War between Athena and Sparta. It is also during this period that Hippocrates thrived.

During the third division, the Hellenistic period from 320-145 B.C., the Greeks power and culture expanded throughout the Mediterranean area. It began with the death of Alexander and ended with the Roman conquest. Alexandrian medicine dominated this period. The last part of the Greek saga is the Roman era that ended in 500 A.D. The overwhelming role that Greek medicine played in the Roman Empire and by the Greek physician Galen is highlighted during this time.

It is beyond the scope of this book to describe in any detail the political aspects, social structure, economy, culture, art, architecture and philosophy of the ancient Greeks except as it relates directly to the development of medicine. Suffice to note that this energetic group of ancient people took important and radical steps to advance a variety of fields. The results of their efforts helped established the very foundation of Western civilization.

One wonders why were the Greeks so extraordinary in many different venues, but particularly medicine? Some medical historians believed it occurred because the Greeks showed intellectual and physical "hybrid vigor." Unlike similar ancient civilizations, the political divisions that prevailed in Greek history prevented the development of a strong central religious bureaucracy. If present, that group's influence would have likely suppressed individualism and critical thought in medicine and other fields. Nevertheless, this lack of theological leadership did not completely prevent religion from having a significant impact on medicine.

Beyond the ecclesiastical aspect, the Greek people were extremely open minded and had an intrinsic willingness to explore new ideas. Their travels and trading throughout the Mediterranean fostered exposure to a wide variety of cultures. People of many professions were permitted to examine current practices in a variety of areas and offer suggestions for improvement. The Greek citizens had an inherent desire to understand the world rationally and reflect upon it aesthetically. Greece provided a rich pasture for nurturing medicine, and as a result, the field expanded into new realms.

Medicine in the Early Archaic Period

In the beginning theological medicine was part of the Greek culture. For many, especially the poor and incurables, medicine retained its religious orientation throughout history. Mythology and the supernatural pervaded early on like in Egypt. The Greeks had many gods who took a hand in producing and curing illnesses. From the start Apollo was regarded as the god of disease and prophecy. In Homer's Iliad it was Apollo who was responsible for one of the most famous diseases in Greek writing—the plague that afflicted the Greek army.

The best-known healing god of antiquity was Asclepius. He was Apollo's son from whom he was thought to have inherited his god like ability to heal the ill. Asclepius taught the Greek philosophy of healing: "First the word, then the herb, and lastly the knife." Asclepius did not just cure the sick, but in some versions of his myth, he also raised the dead back to life. As time went on the worship and legend of Asclepius grew. This was particularly true when the plague arrived. There were not enough doctors to treat all who fell ill. Many could not afford treatment so worship of Asclepius offered hope to those who had no other options. If they survived, it confirmed his heroic nature.

Asclepius had two sons who became doctors. They were known as Asclepiads—physicians from whom all other doctors were said to descend. In those times, the term Asclepiads sometimes also referred to guilds or families of physicians.

The names of Asclepius' two daughters also have a medical connection. They provided the linguistic root for two health related words in English today: Hygeia from which hygiene is derived and Panacea that is the same word in English meaning a cure-all.

Asclepius's reputation flourished over the entire Mediterranean region in the form of temples built to honor his name. These structures were evident well into the Christian era. People came to make offerings to him in the hope of a cure or in gratitude for having received one. In these temples the patient was treated by "incubation." The afflicted person slept a night in the temple and during the darkness god would appear and prescribe a cure. One inscription chronicles this form of healing and demonstrates how close medicine and religion were during the archaic period of Greek history.

"Erasipppe from Kaphy: She slept in the Temple and dreamt that God applied massage to her stomach and kissed her and gave her a cup containing a drug. He commanded her to drink it and then to vomit. She did so and soiled her garment. When she woke up in the morning she saw her garment full of the filth she had vomited, and thereupon she recovered."

A more lasting tradition attributed to Asclepius is the caduceus—the iconic symbol of medicine. Asclepius is depicted holding a long-winged staff that has a snake intertwined around it. The snake was thought to have provided him with a precious healing herb. It was also associated with immortality because the serpent could renew itself by shedding its skin. Some accounts of temple healing involve a snake licking parts of a patient's body.

Thinking how the human body worked was not just a concern of myth and religion. Of decisive importance for the development of Greek medicine was the mutual influence of philosophy and medicine. Natural philosophy provided by the Greeks was one of the great landmarks in the evolution of human thought. A long list of philosophers contributed to their civilization's approach to medicine. All these influential thinkers had an insatiable curiosity about life and the world in which they lived. They formed separate schools of thought that sometimes were in agreement and on other occasions in dispute. A uniquely important aspect of the Greek philosophers' intellectual speculations was that their cognitive reasoning was continuously submitted to criticism rather than being frozen into religious dogmas as would have occurred in earlier times.

They established doctrines and constructed theories including those on the art of healing. Medicine did not represent an intellectually isolated discipline. A person's health posed ordinary everyday problems for these philosophers to consider and seek potential answers. Doctors did not remain indifferent to their ideas. Their teachings became the foundation of Western philosophy and ethics that have had a lasting influence on the world today.

Early pre-Socratic philosophers in the 5th and 6th century B.C. discussed the constituents of the universe, the basic building blocks out of which everything was made. Ancient people were always looking for ways to explain the world. Empedocles (490-430 B.C.) introduced the theory of the balance of four elements: air, fire, water and earth. This was a way to explain the universe that later became a theme of medicine. Recall that the Chinese completely independent of the Greeks used 5 elements (wood, fire, earth, metal, and water).

The concept of "nature in balance" became an important subset of the Greek culture. Hot & dry and wet & cold were eventually added, as were 4 seasons, 4 ages, and 4 temperaments. Hippocrates used the 4 humors (blood, yellow bile, black bile and phlegm) to explain why people became ill. The theory of the 4 humors became imbedded into western medicine for 1500 years until a German physician in 1858 replaced it with his theory of cellular pathology.

Socrates (470-399 B.C.) was the first to develop a method involving an orderly series of questions for approaching any type of problem. This Socratic Method eventually led to the development of a scientific system of inquiry. Plato (427-347 B.C.) learned from Socrates and documented some of his teachings. The philosopher who had the most significant influence was Plato's pupil Aristotle (384-322 B.C.). In addition to be a philosopher, he was also a physician and biologist. Aristotle was a proponent of methodical observation that became the scientific method. His work influenced medicine and science for 2000 years.

Beyond his philosophical writings impact on medicine, Aristotle had a practical contribution. He was one of the first Greeks to understand the importance of anatomy. His followers embarked on a major effort to use animal dissections to comprehend biological matters. He laid the foundation that two millenniums later allowed Harvey to become the first person to correctly understand the blood's circulation.

Hippocrates—"Father of Medicine"

Greek medicine was significantly impacted by Aristotle and other philosophers. Yet, their main effect was not due to abstract theories but to the practical efforts in the field of clinical observation that is best represented by Hippocrates, "Father of Medicine." He is the symbol of the first creative period of Greek medicine. To a certain extent his name alone has come to portray the beauty, value, and dignity of medicine of all times. He and his followers brought a new understanding of medicine by introducing the concept that diseases came from natural causes while at the same time rejecting the role of prayer as the prime method of treatment. Beyond the science, professional dignity for doctors was established by setting a code of conduct in the widely known Hippocratic Oath. Hippocratic medicine flourished in the resplendent period between 480 B.C. from the battle of Salamis that stopped the Persian invasion to the start of the Peloponnesian War in 431 B.C.

Irrespective of his legendary existence very little is known about Hippocrates' personal life. There are no contemporary descriptions of him so medical historians are left with a great deal of speculation. He was born on the island of Cos in 460 B.C. and according to tradition he died at the age of over one hundred. He lived through the period of the building of the Parthenon, the great plague of Athens, the fall of Athens to the Spartans, and the last years of Socrates. After his death in 378 B.C. new legends grew and most of the tales about him were probably exaggerated or untrue.

Writers in antiquity created a whole fictitious genealogy for him. He was said to have been the sixty-second descendant of Asclepius in a direct line back to Apollo. His sons, son-in-law, and grandsons allegedly treated all the princes of the ancient world including Alexander the Great. From about 200 B.C. onward writers made up a set of fake letters in which Hippocrates allegedly corresponds with many other historic figures. These missives advanced different apocryphal adventures about Hippocrates. They became particularly popular later in the Roman period. One historian says that it would be possible to argue that Hippocrates became the "father of medicine" largely by default since he is simply the earliest Greek doctor that there is any available information at all.

Most authors believe that there was a very real brilliant well-known physician called Hippocrates. His famous identity was then conveniently used later on as the title for the collection of sixty or so texts by many different writers that collectively became the "Hippocratic corpus." These early treatises were brought together in the library at Alexandria, Egypt. They serve as the reference point for the Greek's grand theories about the origin of diseases, recipes for remedies, and professional ethics.

Hippocrates himself may have written some of the treatises, but his fame and myth grew because of what was penned by others and often incorrectly attributed to him. The "Hippocratic corpus" is an excellent historic construct of medicine of the period rather than the effort of one man who receives all the credit.

The portfolio was a potpourri that included textbooks, lectures, research notes and philosophical essays. There were many contradictory statements within the various works that were very uneven in value to the reader. But collectively they unite the essence of Hippocratic medicine and the belief that the same physical laws of the cosmos also rule humans. Health and disease was expressed via empirical and rational reasoning that rejected any religious connotation. A few samples from some of the documents in the "Hippocratic corpus" establishes how advanced the thinking had become.

On Ancient Medicine was an important text in which the healing component of the natural world is the primary thesis. The job of the physician was to work with nature to create a harmonious balance. Dietetic observations and solutions were frequently a part of the process. This book also cites recognizable symptoms of certain diseases and observes that climate can affect health and disease.

There is an anatomical and physiological segment in the book titled *On the Nature of Man*. Here fluids in the body were promoted as the main cause of disease not the vengeance of the gods as was thought by virtually all the other ancient civilizations and in Greece's earlier period. These fluids later became known as the 4 humors: yellow bile, black bile, blood and phlegm. This manuscript presents blood as necessary for the life of humans and compares it to the sap of plants. Yellow bile was for digestion. Phlegm was a lubricant and coolant, comprising all colorless secretions including snot, sweat, and tears. It was most evident in illness. Black bile was a constant source of unrest. It colored skin, urine and stool and provoked melancholy.

The humors provoked all manifestations of health. Because the 4 variables created infinite permutations to maintain the body in or out of balance, the theory was difficult to disprove. As a result, humoral medicine would persist as gospel for almost 2000 years in one form or another. From this, multiple cures were developed to recalibrate a humoral deficit or excess. As one author said, "the appetite for a treatment was satiated by the lure of the cure."

The Aphorisms and Oath of Hippocrates

In the book titled *Aphorisms*, Hippocrates is credited with many wise comments about the practice of medicine. These include:

- *Primum non-nocere*. First, do no harm.

- "Life is short, art long, opportunity fleeting, experiment treacherous, and decisions difficult."
- The art of medicine has three factors: the disease, the patient, and the physician.
- What drugs will not cure, the knife will; what the knife will not cure, the cautery will; what the cautery will not cure must be considered incurable.
- Fat persons are more exposed to death than the slender.
- Those who swoon frequently and without apparent cause are liable to die suddenly.

Beyond these expressions, if one were to ask a knowledgeable person to name one thing they know about Hippocrates, many would likely say his oath for doctors. The Hippocratic Oath is the most famous pillar of medical ethics. Like so many other historically influential documents attributed to Hippocrates, it was probably not drawn up by him personally. Nevertheless, it stands as the only Greek medical ethical document of its kind that formally defined the proper conduct of the physician-patient relationship.

The medical profession from the early Greek era to modern times has always had a special closeness to the oath beyond its moral value. The very word "profession" comes from "profess" which in its original meaning was "to swear an oath."

The prototype Hippocratic Oath was not written as a treatise on medical ethics, but rather largely as a response by its author or authors to the free and unrestricted marketplace that Greek medicine enjoyed. Its construction was very likely the result of a perceived need by a handful of Greek physicians to fill a void that had been created by the absence of any legal or ethical constraints on the practice of medicine. There was no professional licensing nor other means to prove who was a competent healer. Anyone could hangout a shingle and proclaim to be a doctor and many did.

The importance of the Hippocratic Oath to the well-trained establishment physicians of early Greece may also reflect the competitive social context in which ancient medicine found itself. In this environment, those who took the oath and voluntarily adopted it could claim authority over other less qualified health providers and justify charging higher fees.

So, what is in this oath that graduates from many medical schools still recite today? The contemporary version is a revised and more politically correct rendition than the initial one found in the Hippocrates corpus. The original pledge consisted of two parts. The first section affirmed the obligation of the doctor to transmit medical knowledge to others, and the second was a summary of medical ethics as a self-regulated discipline.

The Hippocratic doctor swore to the Gods of Apollo, Aesculapius, and other deities to:

- Respect his teacher and his apprentices—this encourages fellowship and the idea of mutual aid in a professional group that obligates the holders of knowledge and techniques to impart this to future generations.

- Keep the secrets the doctor hears in the course of one's work with a patient—this is an elementary right of the patient that the doctor must respect.
- Refrain from sexual relationships with people he/she meets, whether male or female, free or slave.
- Swear not to give poison—this clause has been used in modern times as an argument against physician assisted euthanasia.
- Do not give a pessary to cause an abortion—this does not mean that Hippocratic medicine banned all forms of abortion. In a different Hippocratic text, a slave prostitute is encouraged to jump up and down to cause one.

In the broader context of the pan-Mediterranean world, the Hippocratic Oath is unique and without parallel or precedent in other ancient civilizations like Egypt and Babylonia. None of these cultures produced such detailed oath, prayer, treatise or document of any sort devoted solely to the ethics of medicine.

While its original intent was to secure financial gains for the early Greek doctors may have been less than honorable, the Hippocratic Oath stands alone in its importance to the practice of medicine over multiple generations. Its relevance may have unfortunately diminished in the eyes of some modern medical educators. But most physicians today still believe the ethical principles' it espouses to be as applicable for the contemporary doctor as it was for the practitioners who followed Hippocrates and walked the streets of ancient Athens.

Doctors and Treatments

From the writing of Hippocrates' followers, one can glean a picture of the rational approach of medicine that emerged as well as those individuals that practiced the craft. Iatros was the name for a Greek physician. At the time of Hippocrates, doctors were few in number. The common Greek medical practitioners were basically traveling craftsmen. They had some background in philosophy and rhetoric that served as a foundation for their limited medical knowledge.

These wandering care givers settled temporarily in a town where they entered into the service of a rich merchant or politician. Alternatively, some practitioners worked on behalf of a city council. For most of the time in this capacity, they cared for the poor of the town, the slaves, and intervened when there was an epidemic, war or earthquake. Outside the urban centers, some doctor's practices were not full-time. Their income was supplemented from other sources like land ownership. But the majority of medical providers made their living by delivering their services to groups or individuals, much like the medical system today.

The Iatros doctor was first and foremost patient-centric. This approach emphasized paying close attention to the symptoms of the ill. A physician's task was first to obtain a profile of the patient being treated. This included where they lived, worked, and what they ate. The doctor also examined not only the urine of the patient but anything else coming out of the body in order to determine what was happening inside.

The physician listened as well as observed and touched. For example, in the case of a possible pulmonary inflammation, the doctor shook the patient and listened to the sounds. This was actually not a bad diagnostic technique to evaluate the lungs at the time that was many centuries before the invention of the stethoscope.

The Greeks kept careful records about various afflictions they encountered within their practices. Diseases were classified not only as acute or chronic, but also whether epidemic or endemic in nature. Doctors were primarily interested in prognosis and treatment, not just diagnosis. The concern was for the body as a whole. The doctor's ability to evaluate the course of an illness was critical.

The Hippocratic physicians used the strategy of predicting the future course of a disease in order to gain the trust and confidence of the patient and the faith of his or her family. The doctors' overriding concern was that they could not risk tarnishing their own professional reputation by having one of their patients die while under their care. The stigma of failure to cure often lead to having a physician prematurely leaves the town in which he practiced.

"Our natures are the physicians of our disease" is attributed to Hippocrates. It summarizes his approach to medical treatment. Hippocrates' followers believed they should aid healing, but then let nature take its course. The prime emphasis was to reinstate the balance of humors by treating the entire patient, not just the disease or one body part.

The physicians' therapeutic recommendations to assist nature in re-establish humoral balance included rest, moderate exercise, dietary improvement and a period of starvation to cleanse the system. Various types of bleeding were considered helpful to restore some imbalance. Drugs to cause vomiting or purging were common. While Hippocrates preferred mild remedies, others used an array of stronger nostrums.

If a patient required surgery, it would be difficult to find a surgeon in ancient Greece. Surgery was considered beneath the work of the regular physician. It was either left to others or practiced as a last resort. Though the supporters of Hippocrates did not perform surgery, it interested them. They studied and wrote about it in the Corpus. It included a wounds' treatise that advised realigning fractures, a section on extracting nasal polyps and a part that promoted cautery with a red-hot iron to treat hemorrhoids.

Because of their lack of aggressive medical and surgical treatments, one author aptly stated, "The Greek physicians were, by and large, doctors who sat with patients in one-to-one consultations, deliberating over the kilter of their humors."

The care of pregnant women was limited. At that time, the birth process presented major risks, and women commonly died in labor. Doctors understood the importance of the positioning of the baby and how it was key to a successful delivery. Much was written on how to maneuver the fetus into a helpful position, but it is difficult to know whether this theoretical knowledge was used in actual practice. Cesarean sections were described, but their use was only in the context of removing the fetus from a mother who had died.

Midwives not doctors usually handled women's obstetrical issues. They were seen as reassuring and sympathetic individuals who were free from superstition and not greedy for money. In the larger cities these women would have had basic literacy and some knowledge of the theory behind their medical actions. This was clearly the ideal, but this level of competence was unlikely to have reached outside the urban setting. In the rural regions, the normal midwife would have helped at a birth only occasionally and when available from her regular job. One midwife in late antiquity is reported to have combined her medical role with that of being a barmaid.

Athens Plague

From 430-426 B.C. Athens suffered two outbreaks of a mysterious and often fatal condition that continues to interest medical writers today. The spread of this enigmatic disease began in Ethiopia then passed through Egypt and arrived in the Greek world. It was an exceedingly virulent infection. Sufferers had fevers, headaches, chest pain, and vomiting and violent spasms. Those who did not die lost fingers, toes, and eyes as well as suffering from severe mental confusion. Death arrived usually in seven to ten days.

There is much speculation by modern historians as to whether the culprit was measles, smallpox, typhus, anthrax, influenza or even syphilis. All these contagions have different proponents that believe one of these was the primary cause of this plague in Athens. Most authorities agree that it was not the bubonic plague that struck Europe in the 14th century. No matter what its etiology, the outcome was devastating. It killed nearly 25% of the Athenian soldiers and civilian population. With this widespread calamity the dream of a Greek empire also died.

The sole contemporary source for describing the Athenian plague was by the non-physician Thucydides. He claimed to have suffered from the condition but survived. He undertook the task of delineating the symptoms and the consequences of this outbreak so that future generations would be able to recognize it if it were to occur again. Thucydides also pro-vided a detailed overview of the social disintegration that was a difficult part of the aftermath of the epidemic. He wrote that initially no one wanted to treat the suffering because of the fear of becoming ill themselves. Eventually observant Greeks found that the body-built immunity to the disease, so those that recovered were sometime willing to assist the ill.

Medical historians reading Thucydides' detailed account have suggested that although not a doctor he was the first writer to understand the correct underlying force responsible for a widespread contagion. This is the theory that disease can spread from person to person. It was not until the 19th century that the more scientifically accurate contagion theory replaced its long-standing main rival, the "miasma theory." That hypothesis incorrectly held that the air spread poisons from rotting materials and stagnant water to produce rampant disease.

The Athens plague demonstrated that there were too few physicians with too little knowledge to affect the outcome of the crisis. Hippocrates himself was said to have counseled patients to: "*cito, longe, tarde*"—this translates to "go fast, go far, and return slowly."

The dearth of beneficial medical treatments led to the need for alternative healing. The Greeks once again returned to prayer. They beseeched the Greek God Asclepius to help their loved ones. Many infected people made pilgrimages to the nearest temples named after him. There a cure that advocated fasting, self-mutilation, or taking hallucinogenic drugs was exploited. The fortunate that recovered were expected to leave offerings. Those that survived must have imagined they were leading a charmed life in which no disease could kill them.

Thucydides concluding comment on the plague was quite cogent and as applicable to modern times as in Ancient Greece: "A medical catastrophe can threaten even the mightiest state and no human society should under estimate the speed in which a natural disaster can destroy."

Medicine in Alexandria

The time in which Hippocratic medicine was founded saw the Greek territories expand under Alexander the Great who was one of the greatest military commanders in history. He conquered most of the world that was known to the Greeks at that time. The next era, the Hellenistic period, began with the death of Alexander in 323 B.C. and ended with the Roman conquest in 145 B.C. The center of Greek civilization and medicine shifted from the old Greek settlements toward the new Egyptian city of Alexandria founded by Alexander the Great in 331 B.C.

After the death of Alexander, his large empire was split up with Egypt being ruled by one of his generals, Ptolemy. He founded a royal dynasty at the mouth of the Nile. Under Ptolemy and his successors, Egypt remained politically stable with the Greek elite ruling the local people. The Ptolemaic rulers invested in the arts, sciences, and founded a library that housed books from throughout the known world. The famous library at Alexandria grew until it contained over 700,000 manuscripts of which many works were devoted to medicine. The library lasted for several centuries. It was partially destroyed by a fire during the political turmoil when Caesar arrived in 48 A.D. Its final destruction came with the Muslim conquest 500 years later.

The Ptolemaic leaders' patronage attracted philosophers, mathematicians, astronomers, poets, and physicians from all over the Greek-speaking territories to Alexandria. From this strange cultural melting pot, the most important medical center in the ancient world emerged. Its positive reputation continued even after the fall of the Roman Empire.

Much of the success in the field of medicine was attributable to two individuals who became symbols of the Alexandrian philosophy of medical care. They were Herophilus and Erasistratus. Though not contemporaries, they were often linked together in discussions because of the exploratory paths they both chose to follow. Their teachings and influence lasted until the second century A.D. While both wrote extensively, an accurate picture of these two trailblazers is difficult to obtain because no complete work by either man has survived. What is known about their achievements is based upon later writers who quoted and summarizing their ideas.

Herophilus was born in 335 B.C. and died in 280 B.C. He made important contributions to all fields of anatomy. His portfolio on the brain and nerves is still famous because he correctly saw the brain as the center of the nervous system and the base for intellect. He studied the organs of reproduction. His work revealed many structures not previously known including the spermatic duct, the ovaries, and fallopian tubes. He differentiated between tendons and nerves. He discovered the prostate gland. While he identified these structures, he did not know their specific functions. Herophilus gave the first detailed description of the human liver. In his extensive reflection on a person's pulse, he used a water clock to study pulse rate. He further described the pulses' variations in its size, strength, and rhythm by an analogy to music. He wrote that the pulse is "music in our arteries."

As a therapist, Herophilus had more confidence in drugs, venesection, and surgery than those from the Hippocratic school. His medical thinking is well summarized in one of his aphorisms: "The best physician is the one who is able to differentiate the possible from the impossible."

Herophilus' historically younger counterpart Erasistratus was a doctor's son born on the island of Cos in 304 B.C. He was reported to have committed suicide because of incurable cancer in 250 B.C. He studied in Athens before going to Alexandria. Erasistratus, like Herophilus, was a great anatomist, but was less of a follower of Hippocrates than Herophilus. Erasistratus studied the valves of the heart and saw it as a pump carrying *pneuma* or breath to the body.

Erasistratus approached the riddle of metabolism by weighing both the intake and excrements of birds and noting the loss of substance between the two. He concluded that the bird must have been giving off matter invisibly. He was also likely the first to note pathological anatomy in his observation that a hardened liver was the cause of ascites. As a therapist, Erasistratus was generally opposed to bleeding and the use of multiple drugs.

These early anatomists created new terms for their discoveries. They utilized existing words of their day and analogies from outside the body. For example, Herophilus named one of the membranes surrounding the fetus in the uterus the *amnion*. This was based on the word for soft lambskin indicating the protective function of the membrane. In studying the eye, he designated one of its membranes 'net-like' from which we get the word 'retina' that is based on the Latin for net. He discovered the duodenum (the first section of the small bowel) that is the Greek name for 12 fingers that is the linear length of that organ. Words like these, still in use today, had a very different feel in their original context.

The discoveries of Herophilus and Erasistratus would not have been made without the opportunities that they were given to study the human body. Up until then human dissection—let alone vivisection—was unknown in the Greek world. It is difficult to explain why this practice was permitted and even encouraged in third century B.C. Alexandria.

The Alexandrian physicians benefited and were influenced by existing Egyptian culture. The Egyptians accepted cutting into the body as part of the embalming process, and this opened Greek minds to the possibility of dissection. But it more likely occurred because Alexandria was a frontier city. There was no retribution when new ideas and concepts were considered. In practical terms, it is likely that the patronage of the Ptolemaic rulers allowed and tolerated behavior that would have been impossible in classic Greece. Some medical historians have even suggested that criminals and prisoners of war were supplied from the royal jails as victims for vivisection. Though the practice of dissection was sanctioned, it was still controversial. Some leaders of the day felt that anatomical research distracted physicians from their care of patients.

After this time, Western medicine abandoned the dissection of humans for over 1500 years. A different medical philosophy developed in which dissection was seen as scientifically unnecessary because dead bodies could not reveal anything about living people.

Modern medicine owes a great deal to the success and advancements of the ancient Greeks. The foundation laid by Hippocrates and the innovations in Alexandria form a lasting legacy of achievements that is greater than the sum of all the individual parts.

CHAPTER SEVEN
Roman Medicine

The stage was set for the last great segment of medical care in ancient Greece. It was the assimilation of Greek medicine's accomplishments into the Latin speaking world of central Italy, and subsequently over time into all of Western Europe. This represents one of the most momentous developments in the history of medicine.

A collection of systems of medicine in one society was transplanted into another with a different language, culture, and political structure. Without this progression it is possible that Greek medicine would have remained on the same lower level of importance as the Babylonians and Egyptians. Greek medical theories continued to be studied, applied, challenged and defended in Western Europe well into the 19th century through their Latin versions and texts.

Beginning in the third century B.C., Greek physicians, free-man, and slaves slowly migrated into Rome. Initially, the Roman authorities opposed the use of foreign physicians in their territory. But because the accomplishments of the native Roman population in the field of medicine were conspicuously absent, Greek physicians were soon embraced.

Doctors in Rome

In the late second century B.C. the permanent establishment of Greek medicine in Rome was largely attained by the strong personality of the Greek doctor Asclepiades. Named after Asclepius, the healing god of antiquity, Asclepiades was born in 124 B.C. and died in 40 B.C. He believed that humans were built of invisible particles or atoms, and that health was a function of their free and balanced motion in the body. Disease was due to a mechanical disturbance of the movement of these atoms.

What gained Asclepiades' reputation in Rome was less his theories than his therapies. His slogan for treatments was: *"cito, tuto, jucude"* or "swiftly, safely, pleasantly." This was an adaptation of ideas in the Hippocratic corpus. His liberal use of wine and gentle exercise was notorious among his opponents, but at the same time it was very enjoyable for his patients.

He was famous for championing five types of basic therapy:
- Regulating the intake of food and wine
- Massage

- Ambulatory exercise
- Rocking gentle calisthenics on a swing, to provide passive workout for those unable to indulge in more strenuous activity
- Bathing, which was passive hydrotherapy

He also used powerful drugs, and clysters or enemas to purge the bowel.

In Asclepiades, Greek medicine in Rome found its earliest celebrated advocate. His example undoubtedly encouraged others to come to Rome to seek their fortune. His activity marks the time that Greek medicine was effectively transferred to Rome. Henceforth, in spite of a different language, a different theoretical emphasis, a different legal and different social structures, medicine in the Greek and Roman worlds came together as part of the same intellectual universe.

Several medical sects subsequently developed. The most dominant was Methodism founded in 50 B.C. It was called this because as the name implies, it followed a uniquely simple theory of medicine and confined therapy to a few successful methods of healing. Disease was caused by either a narrowing of internal bodily pores or by excessive relaxation of these pores. Treatment concentrated on reversing the process. Its genesis is thought by some to be because of the necessity of treating large numbers of patients in the slave population with a minimum of effort.

Among the most influential rivals of the Methodists were the Pneumatists. They were called this because this group placed a great emphasis on *pneuma* or spirit as the controlling factor in health and disease. Their theories were a mixture of stoicism and Hippocratic medicine. They paid particular attention to the environment, the seasons, the layout of towns, and the design of houses as a contributing factor to maintain good *pneuma* and a healthy balance in the body.

Notwithstanding their diversity, the Pneumatists, Methodists and Empiricists (a group that was started in Alexandria but came to Rome) all held agreement on one thing. Hippocrates and his writings were worthy of great attention and respect and would remain so until Galen came to prominence. Before that occurred some of the pure Roman aspects of medicine during this period are worth highlighting.

With the exception of a handful of famous elite doctors, the status of physicians in the Roman world was never very high. In fact, early Romans citizens did not enter into the practice of medicine. Doctors were considered craftsmen. Like in earlier Greek times some Roman cities employed public physicians. For the less fortunate Roman citizens who could not afford to pay a doctor's fee, medical care was performed within the family based upon the authority of the *paterfamilias*, the male head of household. Roman residents had an idealized sense of self-sufficiency and thought that the best medicine was very simple and as a result used home remedies. A good diet and specific foods were encouraged to fend off illness. The famous Roman politician Cato believed that cabbage was the key to good health. He lived to be 84, an exceptionally long life for that time.

The Romans were not above seeking out religious cures. In 294 B.C. the citizens of Rome encountered a terrible plague. Herbal remedies were ineffective. In desperation, they turned to the old Greek God of health, Asclepius. When a contingency of Romans visited an Asclepius temple in Epidaurus, a small city in Greece, it was reported that a snake was seen boarding the Roman ship. The snake, as previously noted, was the Greek symbol for rebirth. The Romans interpreted this as a sign that the Greek god of medicine would help Rome overcome the plague. Coincidentally, the epidemic lessened shortly thereafter. As a consequence, Asclepius was worshiped like in ancient Greece in magnificent temples built throughout the Roman Empire.

Public Health

Beyond the occasional individual Roman physician who was not of Greek origin, the primary contribution that the Roman establishment achieved in medicine was in the field of public health. While the Greeks had valued art, philosophy, and science, the Romans emphasized the utilitarian.

The concept of creating an optimum environment for good health was integrated into most aspects of Roman life. The various rulers focused on engineering accomplishments and made a commitment to establish a superior public health program to serve people of all income levels. They built aqueducts for water delivery, created public baths to improve cleanliness, and generated methods of community waste removal not only in Europe but everywhere they went.

Hospitals were originally built for the military to treat the injured as they conquered other nation states throughout the world. They were permanent well-designed structures. Some had internal heating and all were well ventilated. While they did not understand about germs or contagion, the Romans realized that being around a person who was ill sometimes made others sick. This lead to establishing separate patient rooms that reduced exposure of diseases between those that were hospitalized. Outside the military, a wealthy Roman matron, Fabiola, founded the first civilian public hospital in Rome in 390 B.C.

The native Roman doctors did not advance any new theories of disease, provide legendary treatments, or have famous physicians like the Greeks living among them. But they did have government sponsored engineering accomplishments that spearheaded a superior public health program. It would not be replicated again for many centuries.

Galen of Pergamum

It is impossible to think of medicine in the Roman Empire without Galen of Pergamum who lived 400 years after Hippocrates. Galen was the most influential and prolific writer of all physicians in antiquity. Fortunately, much of his material has survived. He brought together earlier medical conjectures about the body and merged them into a synthesis that was all his own. He became a pivotal figure in the history of Western medicine where his theories dominating health care for 1,500 years.

Galen was born in 129 A.D. into a wealthy family that lived in Pergamum—a region that today is in western Turkey. The area was at its height of prosperity within the Roman Empire during his lifetime. As the son of a very rich architect, Galen's father had enormous influence on him while growing up. Initially, he studied Greek literature and philosophy. Then as a consequence of a dream his father pushed the 16-year-old Galen towards the study of medicine.

Galen's initial training began in 145 A.D. in his hometown. From there he went to Corinth where he learned from various famous Greek teachers of medicine. Subsequently, he traveled to Alexandria where his studies continued for many years—long past the ordinary time when most physicians would have begun their own practices.

In Alexandria, he acquired knowledge about Hippocratic medicine. Galen worked hard to master the original teachings of the man he thought of as the great Master of Healthcare. Unlike the earlier Greek tradition, he studied anatomy and surgery. Galen was known to be meticulous in taking notes and integrating his medical education with his earlier studies of philosophy. He maintained this distinctive intellectual process throughout his adult life. As part of his scholarly pursuits, he traveled to India and Africa. There he mastered pharmaceutical remedies that originated from the native plants in those parts of the world.

In 157 A.D., Galen returned to his home in Pergamum 12 years after leaving. He became the physician to the gladiators. He claimed to have significantly reduced the death rate among those who were wounded. After four years of healing these injured fighters, Galen departed for Rome. There he soon gained fame as a practitioner, lecturer, and experimenter. After briefly leaving the city because of a plague, he returned back to Rome where he became the court physician to emperor Marcus Aurelius. After curing him from an illness that two other doctors failed to remedy, Galen's reputation was assured. The appointment greatly enhanced his fame and fed to his highly egotistic, self-confident and bombastic personality. As one historian notes "Galen's wordy, aggressive, and self-laudatory writings do not reveal a very attractive personality."

Galen's Medical Beliefs

Despite Galen's "self-made man who worships his own creator temperament," his medical beliefs were firmly rooted. He was a champion of Hippocratic teachings that emphasized close observation and an analytic approach to assessing the 4 humors.

Galen taught and succeeding generations of doctors believed that the basic composition of sick persons was an unbalance of 4 humors—blood, phlegm, yellow bile, and black bile. In order to establish better health, they had to be realigned. Blood was the humor of hot and cold. People in this category could either be sanguine or have hopeful personalities. Those that were dominated by this humor had healthy complexions and were cheerful, warm, and generous. Phlegm was categorized as being cold, wet, and slow to anger, but also sluggish, dull, and detached. Yellow bile folks were hot and dry. An excess was thought to make one irritable and in a constant state of agitation. Black bile individuals were gloomy, depressed, melancholy, cowardly and envious of others.

When these 4 humors became asymmetric, illness was the result. For example, too much phlegm caused lung disabilities. So to cure these problems, the physician needed to find ways to rid the body of phlegm. These adjustments were achieved through diet, medicines, and bloodletting. In addition to restoring balance, Galen recommended a system of opposites. Fever was treated with something cold. Weak people were given difficult exercises to build up strength, and those with chest symptoms were told to perform singing exercises. Trial and error was at the heart of all methods used.

Galen differed from Hippocrates in one major aspect of treatment. Hippocrates felt that balancing misaligned humors causing disease needed to be achieved within the whole body. Galen introduced the concept that balance could also be obtained organ by organ. This allowed physicians for the first time to develop organ specific remedies. This change in philosophy became a powerful influence on medical science in the future.

Galen believed that medicine needed both practical and theoretical elements. He championed the notion that anatomy was the foundation of medical knowledge. Because in the Roman world it was taboo to cut into human corpses, he conducted his dissections on animals. He preferred using Barbary apes and domestic pigs. While mistakes were made because the anatomy of animals and humans often differ, he identified more about anatomy than those who preceded him.

Unlike Hippocrates, Galen gave credence to the power of pharmaceutical cures. He sometimes had as many as 25 different drugs in one prescription. Such complicated compounds were called "Galenics." Bleeding and laxatives frequently accompanied these concoctions. His enlightened approach prescribed climatic treatment for tuberculosis. He was concerned for proper hygiene and promoted prevention as being preferable to treatment.

Galen learned to be an adequate surgeon while treating the gladiators in his early years. But indicative of the beginning of the separation of surgery from noninterventional medicine, Galen no longer practiced surgery to any great extent after immigrating to Rome. Galen subscribed to the theory of "laudable pus." This hypothesis stated that every wound normally produced pus in the healing process. As a result of Galen's strong influence, this ineffective traditional theory persisted into the 19th century when the appropriate aseptic treatment of wounds finally emerged.

Galen's impact was monumental in nearly every aspect of medicine. His followers admired his methodical analytic techniques, independent judgment, and cautious empiricism. As a self-promoter of his own genius, his works traveled far. His texts were heavily used and set the medical world on a more positive path that led to the understanding of medicine as a science in spite of his significant overreach in many areas. Galen's influence was so powerful and remained so potent for 10 centuries that scientific thinking became paralyzed and did not progress with new ideas as part of a continuum of medical learning.

With the fall of the Roman Empire in 476 A.D., Western civilization entered a very bleak period. The Greek medical beliefs that had been adopted by the Romans were largely replaced by a reversion to religious healing. Many of the gains made during the first centuries A.D. were to lie dormant for a thousand years.

PART II
MEDICINE IN THE MIDDLE AGES
AND
RENAISSANCE

After the collapse of the Roman Empire in 476 A.D. for the next 1200 years Western Europe underwent monumental alterations. Historically it included two fundamentally different periods: The Middle Ages and the Renaissance. The broad social and cultural patterns during these years were more apparent in the practice of medicine than in any other discipline or skilled occupation. Healthcare during these 12 centuries was principally a narrative of many dissimilar individuals—traditionalists mixed with iconoclasts bent on change. There were men and women, laity and clergy, learned scholastics and the self-educated.

All of them faced the challenges of their times. More often than not, they just adopted past theories and treatments. But there were a fair number that embraced an innovative spirit much like modern medical scientists. Those who dared question the established doctrines were fascinating individuals. Their solutions to problems ranged from the bizarre to the practical and included amazing insight that was well ahead of their times. The revolutionary concepts that they introduced at the end of the Renaissance set up medicine for the scientific changes that occurred in the 18th and 19th centuries.

Background

Before reviewing the medical practices and innovations that occurred during the Middle Ages and Renaissances, knowledge of a few famous non-medical events and people of these two great periods furnishes perspective to the accomplishments of the physicians of the age.

At the conclusion of the fifth century A.D. a coalition of Germanic and other tribes marched through Italy seizing lands and power. The emperor Romulus Augustus was disposed and exiled. This ended the Western Roman Empire and began the drift of Europe into 1000 years of obscure history called the Middle Ages. It closed with the flowering of the Renaissance in about 1500 A.D.

The medieval era has been described as "a thousand years without a bath." Certainly it was a time of ignorance, superstition and violence, but it also was one of great adjustments. There were assorted noteworthy achievements in society that laid the foundation for years to come.

The first half of the Middle Ages from 500 A.D. to 1000 A.D. is often called the Dark Ages. There was a dramatic economic decline during those 5 centuries. European society was reshaped as nobles became kings and ruled over vast regions. This became the feudal system. Out of this period the knights emerged. They pledged their allegiance to kings and other lords and fought to acquire riches. They went on Crusades to take the Holy Lands back from the Muslims who had conquered the eastern area of the Roman Empire in the 7th and 8th centuries.

Throughout these early medieval years, the Catholic church began to strengthen its hold on society. It became the predominant cultural influence in medieval Europe. At the height of its influence, the clergy had significant power over the kings and the governing of everyday affairs of the common people. But in the late Middle Ages the church's moral authority decreased as the hierarchy was tainted with political and financial corruption in its leadership.

The zenith of the Middle Ages was from 1000 A.D. to 1500 A.D. A permanent legacy of that time is visible today in the soaring cathedrals and massive castles that sprang up all over Europe. Society became more complex as a new kind of educational system appeared, the university. This spawned an interchange of ideas and cross-cultural encounters that contributed to the weakening of feudalism. This movement helped bring about more participatory governance best exemplified by the Magna Carta in 1215.

The end of the Middle Ages is hard to pin point, but it is usually dated shortly after the fall of the Byzantine Empire when Constantinople was captured by the Ottoman Empire in 1453. Other historians use Columbus' discovery of America in 1492 as the terminus. Medicine during the 1000-year medieval period struggled to show improvements. Even so in the centuries just before the Renaissance there were more enhancements in healthcare than one would have expected.

The events that made up the cultural movement known as the Renaissance (the word is from the French meaning "rebirth") began in 1500 A.D. and concluded in 1700 A.D. These two centuries marked the opening of the "early modern age" of history.

Modifications did not happen all at once. There was a gradual transition from the late Middle Ages in which a number of factors were decisive in contributing to the Renaissance's formation. These included: the large-scale introduction of gunpowder and its effect on warfare, the invention of the printing press, the discovery of new sea routes for trade and exploration, the introduction of a money-based economy, and finally after the defeat of Constantinople the spread of Greek refugee scholars all over Europe.

Historians say that the Renaissance started in Florence Italy. It subsequently spread throughout Europe where its influence was felt in art, philosophy, music, politics, and to a great degree the very core of medicine. The heart of the movement was reviving the importance of embracing classical Greek and Roman learning as a foundation and stepping stone to explore and question all categories of issues that effected man and his social environment. This approach was revolutionary coming as it did after the Middle Ages when religious dogma, authority at multiple levels and superstition dominated all thinking and discouraged the pursuit of new ideas.

This new outlook was called humanism. It was a novel method of learning. As one historian noted humanists asserted "the genius of man was the unique and extraordinary ability of the human mind." Advancements were made in the theories of mathematics, geography, and natural history. There were inventions related to the fields of engineering, mining, navigation, and the military arts.

Early on in 1543 Copernicus published his sentinel theory that the world was not the center of the universe, but rather the earth and the other planets orbited the sun. He propagated the "Copernican Revolution" that not only reordered astronomy forever but ultimately all of science including medicine. Galileo and Isaac Newton followed with their own breakthroughs ideas.

Outside of the basic sciences in art painters developed the use of perspective and the wider trend towards realism in which there was the desire to depict the beauty of nature and man. This resulted in artists like Leonardo da Vinci, Michelangelo, and Raphael accurate portrayal of the human body in paintings and sculpture. Their work aided in the advancement of medical anatomy.

With the help of the printing press—invented in 1455—the rapid transmission of the new ideas of the Italian Renaissance spread to the other European countries where similar changes occurred. In England this is the time of William Shakespeare and Sir Thomas Moore. In Germany, Martin Luther split from the Roman Catholic church. The Renaissance ended in 1700 that also marked the beginning of the Age of Enlightenment.

CHAPTER EIGHT

The Influence of Christianity and
Islam on Medicine
in the Medieval Period

The 1000 years of the Middle Ages beginning in about 500 A.D. was a gloomy era marked by ignorance, brutality, and intellectual decadence. The medical history of the Dark Ages (the name sometimes given to the first 500 years) is a difficult age to understand properly.

The challenge for those who practiced medicine at that time was to weld together the pagan ideas of the invading barbarian tribes from the north with the classical tradition of the defeated Roman empire and the Christian religion. Early medieval medicine included features of all three sources in varying degrees. With the passage of time there was an amalgamation of these elements that brought a measure of progress.

Because medical treatments were so ineffective in the early Middle Ages, many sought spiritual intervention by the church. The clergy was not totally intolerant of secular medicine. One should not be totally surprised at this trend considering that Jesus, the founder of Christianity, in his ministry healed the sick and cared for the ailing. Also, one of his disciples, St. Luke, was a physician. The early Christian religion shared common ground with medicine in another way. Etymologically, the words 'holiness' and 'healing' stem from a single root, conveying the idea of wholeness.

Christian ideas about medical science set boundaries between the body and soul implying the subordination of medicine to religion. For the church the doctor was the person present merely to treat the body while the priest was there to cure the soul. This demarcation between temporal and eternal often became blurred but generally enjoyed a peaceful coexistence. As a consequence, in the dominant Christian society of the early medieval western Europe there was no abandonment of ancient medical knowledge or the disappearance of secular medical practitioners. They existed side by side but under the arm of the church.

For the early theologians, in the most general terms, sickness like all other evils afflicting human life was conceived as an outcome of sin, possession by the devil, or the result of witchcraft. Healing was obtained through prayer, penitence, and from the assistance of the saints. Every cure under these religious circumstances was regarded as a miracle by the church.

Religious doctrine did not exclude the idea that there were also prevailing secular based causes of disease, nor did the church prohibit endeavors to restore physical health by both standard and supernatural means concurrently. So as time passed, medical knowledge and healing activity tended to come more and more a part of and within the orbit of different ecclesiastical communities. This allowed medicine that embraced the historic Greco-Roman theories to survive initially in a clerical religious or monastic environment.

Monasteries

The monks in the monasteries played a predominate role in both the practice of medicine and the composition of medical texts. This seems to validate that Christianity did not hamper medical development directly because the study and writings about health in the medieval era was not completely prohibited. The work of the monks was primarily of translations. This brought about some amalgamation of the scientific point of view with that of Christianity. But the church authorities left no room for independent scholarship or challenging of old theories. In spite of that restriction there were practical treatises that were helpful in maintaining the cloister infirmaries and herb gardens. The monks were particularly interested in healthcare for the sake of the brethren themselves, but also so they could be beneficial to the villages residents that gathered around the monasteries.

The Council of Clermont officially brought the period of monastic medicine to an end in 1130. It forbade healthcare practices by monks because it was considered to be too disruptive to an orderly life in monastic sequestration. Medicine did not automatically become a layman's occupation, but rather it fell into the hands of the secular clergy.

St. Hildegard of Bingen

Medicine and religion was not just intertwined with the monks in the monasteries but it was also present in the convents. This is best illustrated by the life of the Rhineland Benedictine abbess known as St. Hildebrand (1098-1179). She was also referred to as Hildegard of Bingen.

Her life demonstrates that the convent sisters were very much interested in things intellectual more than was initially evident historically until recent research. Hildebrand rose to prominence as a healer and is remembered because she wrote extensively about medical conditions and cures.

She was the tenth child born into a family of free nobles who lived near what is now Frankfurt, Germany. It was the custom of the time for families to "tithe" their tenth child to the church. Ten children were a lot to feed and clothe. This practice helped lighten the family's burden while providing the church with an additional person to work. At age eight Hildegard went for her education to a nearby Benedictine cloister. After her schooling was finished, she became a nun.

Over the decades, she became well known because of her active correspondence with nearly every important man of her generation. In spite of the extensive time taken letter writing, she found the leisure to write a series of books. Two were medically related and composed after she had "divine revelations" about the causes and cures of many diseases. They were written in order to provide information mainly for the nuns who were in charge of the infirmaries of the monasteries of the Benedictines.

The first was called *The Book of Simple Medicine*. It was encyclopedic in its scope. It was marginally helpful medically and also included magic spells. Her advice was wide ranged from what to do about a simple cough to leprosy.

Hildegard preached moderation in all things, exercise, and a good diet. Wholesome spelt—an ancient grain—was a favorite along with boiled hedgehog. Butter was to be avoided along with peacock and falcon. She hated junk food because it "spread slime in the stomach like a rotting manure pile." But she did not begrudge the occasional treat, such as cookies made with gold. She said burping and hiccups were precursors of cancer, bad weather caused postnasal drip, and running too fast could shrink the testicles and send toxic phlegm to the brain.

She advocated the use of plants and herbs, as they were God's gifts. In doing so she achieved great renown for laying the foundation of botanical studies in northern Europe. In fact, one section from her book was recently reissued as a modern manual because of the resurgence of the natural healing philosophy that uses plants and herbs for treatments.

Beyond medicine she was an accomplished composer. Many of her musical works are still known and played today. Hildegard lived at a time when few women were in positions of authority, but she was respected for her work and consulted by bishops, popes, and kings. She was truly a Renaissance woman long before that era began.

The Rise of Islam

While the church and the feudal government system influenced medicine in Western Europe, the eastern half of the old Roman conquests grew and prospered as the Byzantine Empire. It covered a large geographic area including Arabia, Persia, Syria, Egypt, North Africa, and Spain.

Unlike the restrictions in the early Christian church, a culture of scholarship and learning was permitted to grow under Islam. A tolerance for other viewpoints within the Islamic religion at that time was an accepted practice. For 800 years medicine within the Muslin societies spread. It was a new force that impacted the direction of healthcare in the Western world. This time period is often labeled the Golden Age of Islamic medicine.

Islam was founded in the early 7th century by Muhammad (570-632 A.D.). He began his life as a poor orphan but rose to become a wealthy merchant. When he was forty, he received a vision in which the Koran was revealed to him. He gradually assumed the mantle of the last of the prophets in the long line that began with Adam and Noah.

By the time of Muhammad's death, Arabia had been won over for Islam. Under his successors (called Caliphs) his followers eventually conquered almost all the lands that made up the Byzantine Empire. In these Arab countries, Islam was not a proselytizing faith. Christians and Jews were given a special status. This cooperative attitude between Christians and Muslins changed during the Crusades that took place from 1096-1272. What brought some unity to the Arab empire was not religion but a common language.

The initial peaceful era set the stage for the scientific investigations of the period. It began with the assimilation of ancient medical treatises into the Arabic medical community through translating the traditional Greek medical literature into their own language. The Islamic practitioners then added their personal observations and treatments. This led to the Arab physicians creating their own classic medical literature.

Al-Razi

The first great Islamic medical writer and leading physician was a Persian, al-Razi (860-932 A.D.). He did not study medicine until later in life. According to tradition, he supported himself as a singer until he was thirty and became a physician. Like most Arab doctors he was intrigued by the philosophy of Aristotle and thought of Galen as his master.

Similar to other prominent Islamic doctors, he produced multi-volume medical encyclopedias that preserved and organized classical medical knowledge with great thoroughness. He felt that by carefully studying the works of the ancient practitioners, the Arab physician could assimilate the experience of these deceased colleagues as if having lived thousands of years ago.

He was not just a mere compiler and translator of old theories and treatments. Like other Islamic scholars, al-Razi did not blindly accept all the words of Galen and others. He admired them but said that doctors needed to question all medical writings and form opinions for themselves. This approach was radical thinking at this time in the Middle Ages when Church authority was sacrosanct among practitioners in Western Europe.

Because of this unique philosophy, al-Razi became an excellent clinician in his own right. He worked from the fundamental belief that when it came to looking for a cure, good sense and experience offered the ultimate authority. He felt that all that was written in books was often worth less than the experience of a wise doctor. But this ideology did not stop him from writing over 200 texts.

His major contribution to medicine was documenting contemporary case histories. In his best known book, he used this method to discuss diseases beginning at the top of the body at the head and working down to the feet. Elsewhere, al-Razi's famous treatise on smallpox and measles was the first real study of those entities. He recognized them as two separate illnesses. He cautioned physicians not to make a judgment too early in the assessment of their patients because of the initial presenting similarity of both diseases.

Like Hippocrates and Galen, he composed many aphorisms. One commented about the value and the influence of the mind over body in serious organic disease where death seemed impending. He said "Physicians ought to console their patients even if the signs of immediate death seem to be present. For the body of men are dependent upon their spirits."

One story illustrates al-Razi's bravery and creativity. A wealthy patient of his was suffering from a crippling ailment that limited his mobility. Al-Razi was called to his home. But he only came and began his evaluation and treatment after being assured that there was a fast horse readily available to him outside the patient's house.

While the patient was soaking in a hot bath, al-Razi added various healing substances to the water. After administering some additional remedies, the doctor suddenly brandished a knife and starting yelling insults at his patient. He then ran outside and departed on the horse that was waiting for him. The disturbed patient was so angry that he scrambled out of the tub and chased after the departed doctor.

Al-Razi after some time returned. He explained to his patient about his unusual upsetting actions. He said that what he did was all part of his cure to get him up from his crippled state and moving again. The patient forgave him after realizing what the doctor did worked, and he was now walking. He rewarded al-Razi with gifts and money.

Al-Razi also had many other talents. He was a poet, mathematician and chemist. After having won fame, he retired to Bagdad. Unfortunately, he developed glaucoma and became blind. In death, he is remembered because there was scarcely any aspect of medicine that al-Razi did not touch, and he frequently gave better advice than those of future generations of medical providers.

Ibn Sina

Another Islamic physician who demonstrated the depth of knowledge that Arabic medicine had during this period is Ibn Sina (980-1037 A.D.). He was also known as Avicenna. Born 50 years after al-Razi, Ibn Sina is sometimes referred to as the "Prince of Physicians." He mastered the Koran by the age of 10 and went on to study philosophy, natural sciences, and medicine.

Recently, historians have reevaluated his over 270 works. Their reassessment revealed that Ibn Sina made important fundamental discoveries that had to be uncovered centuries later to be appreciated. Among the amazing findings was his description of the elements of Newton's First Law of Motion, a full 600 years before Newton. This axiom is sometimes referred to as the law of inertia: An object at rest stays at rest and an object in motion stays in motion with the same speed and in the same direction unless acted upon by an unbalanced force.

Even more mind-boggling is Ibn Sina descriptive treatise on time and motion that reached the same conclusions Albert Einstein did in his theory of relativity in 1905. This hypothesis says, "all motion must be defined relative to a frame of reference and that space and time are relative rather than absolute concepts."

The importance of his major book on health, *Canon of Medicine*, did not rise to the same level as his ability in foreseeing future events in physics. But with it came an appreciation of Galen and Aristotle that did have enduring value. From 1250-1600 A.D. his book was used as a reference tool both for diagnosis and treatment by doctors in clinical practice and university professors.

His treatise provided instruction on how to inspect urine, evaluate pulse, diagnosis disease symptoms, and administer practical treatments. In Ibn Sina's remedy for joint pain, he recommended soothing olive oil baths. When pain was particularly unbearable because of damp weather, then boiling a lizard in the oil would make the bath more effective. In other discussions he wrote of his concern for water quality, mind and body interactions, and surgery. Ibn Sina's views became the standards for the era.

Outside of his devotion to work, Ibn Sina had quite a animated personality. He was said to have had a great love of wine and women. Over time too much alcohol affected his health, so he began to treat himself with multiple medicated enemas every day. The process caused ulcers, seizures, and extreme weakness and ultimately led to his death likely caused from his own "cure." His behavior certainly validates the old medical axiom that "a doctor who treats himself has a fool for a patient."

Ibn An-Nafis

An example of the willingness to doubt old doctrine is seen in another Arab physician, Ibn an-Nafis (1210-1288). It was not until 1924 that Western science discovered his breakthrough theory about the heart-lung circulation. An-Nafis investigated the anatomy of the heart. Galen centuries earlier had said that blood moves from one side of the heart to the other through invisible channels. But an-Nafis used his own observations to declare that Galen must be wrong. These imperceptible canals in the heart could not be seen because they did not exist. An-Nafis postulated that the blood moved from the heart to the lungs and then returned back to the heart from which it circulated throughout the body. His idea was brilliant and correct, but it took another 300 years before it was validated during the Renaissance.

Medical historians have debated just how an-Nafis conceived this breakthrough hypothesis with only minimal experimentation. Western scholars have concluded that he either rendered a lucky guess or that he deduced his heart-lung circulation conjecture after making the reasonable assumption that the solidity of the cardiac septum made it impermeable. Unfortunately, no one built on an-Nafis' enlightened theory. His influence did not live beyond his time and was subsequently lost to history until the 20th century when his brilliance was finally appreciated.

The Importance of Islamic Physicians

Why is it so important and worth remembering Ibn Sina, al-Razi, Ibn an-Nafis and other Islamic physicians? Their impact was twofold:

- First, through their writings and translations of old classic medicine they preserved for future generations the theories and practices of ancient scholars.

- Second, concurrently and likely more significant, was their willingness on some occasions to question and criticize not only their own work but also the beliefs of the Greeks and Romans. This attitude meant that academic Arab medicine developed at a faster rate than medical science in early western Europe. There doctors were restrained by Church authorities and reluctant to challenge the theories of the old masters like Galen.

Moses Maimondies—Famous Jewish Physician

Not all classical Arabic medicine was developed and promulgated by those of the Islamic creed. Within both the Byzantine

Empire and in Christian European countries, there was a
tolerance of those of the Jewish faith. Jewish physicians
reached distinction at all times during the Middle Ages. Their
influence on culture and promotion of science was extensive.
As a rule, they stood for what was best and highest in education. They became leaders in medieval medicine because the
main challenge of that period was the preservation of the
Greek legacy, and they were one of the best custodians of that
tradition.

The story of Moses Maimonides serves as a prototype of the
many superb Jewish doctors of the Middle Ages. While a
product of the Islamic era, his influence extended well
beyond his time and his geographic Mediterranean home. His
works long after his death affected the great thinkers and
teachers within the flourishing universities of Western
Europe that laid the foundation for a new mold of medical
education

Moses Maimonides (1135-1204) was the most famous Jewish
physician of the Middle Ages. He was born in Muslim Spain
and his work paralleled the Islamic intellectuals of his day.
After he was forced to leave his homeland, he moved to
Cairo where his medical practice made him a celebrity. He
was appointed the court physician to the Sultan and became
head of the Jewish community in Egypt.

Maimonides was a true generalist combining philosophy,
logic, theology, astronomy and medicine. He wrote hundreds
of treatises and books on different intellectual subjects. The

ten that concerned medicine all have survived. They were penned mostly in Arabic and some were in Hebrew. They addressed a variety of topics including hemorrhoids, asthma, and poisons with their antidotes. There was little originality in Maimonides' work, but his intuitiveness, extensive knowledge, and discriminating wide capacity made his writings of special value.

His medical aphorisms and letters set rules for life and death. These idioms became part of popular common sense medical traditions of the era. Some examples include:

(1) A man must not overload his stomach but be content always with something less than necessary to make him feel quite satisfied.

(2) Men should sleep eight hours and leave their beds when the sun rises.

(3) During sleep one should not lie neither on his face nor back, but only on his sides.

(4) Honey and wine are not good for children, though they are beneficial for older people.

(5) As long as a man is able to be active and vigorous, does not eat until he is over-full, and does not suffer from constipation, he is not liable to disease.

While not adding anything cutting-edge to Arabic medicine, Maimonides' considerable literary output earned him respect. Like other medical practitioners in the Byzantine Empire he was widely cited by European authorities that profited from his writings.

Medicine for the Arab People

While the scholarly physicians within the Arab nations wrote about the practice of medicine, healthcare for the people was not unlike that in Europe during the early Middle Ages. God as punishment sent disease, and evil spirits caused illness. If sickness was to be treated, folk remedies supplemented by magic were the methods of choice that prevailed.

As independent Islamic scholars and doctors became more interested in Greek medicine, they needed to reconcile their practices with the beliefs of some who felt that religion and prayer should outweigh all else. Those who argued for the Greek style of medicine advanced that medical care was a form of service to man. They persuaded others that the art and practice of any type of healing was second only to faith in earning God's blessing.

To serve the sick, a range of medical practitioners and services were offered. There were various ways to prepare for a medical career. Some doctors were self-educated while others underwent formal study under a teacher. Muslins often taught medicine in mosques. Instruction in the hospital setting was common since patients were on hand and many of these institutions had libraries. While the curriculum varied, most focused on the key works of Galen. Once training was completed, medical practices were unregulated with no licensing requirements or laws defining the profession. Outside the cities learned medicine was unavailable. For the sick in this setting their care was provided be a spectrum of healers who dispensed popular remedies. Many of them had several occupations. Sidelines in trade were common.

The traditionally trained Islamic doctor classically wore a white shirt and cloak, a distinctive turban, and carried a silver headed cane. Generally, the physician was also perfumed with rose water, camphor, and sandalwood. Arab medical practitioners were broadly successful financially because they believed in charging higher fees to the wealthy so that they could afford to treat the poor without charge.

Women were permitted to train usually in the private setting with a tutor as nurses, midwives, and gynecologists. Since females were not allowed to be treated by male doctors, women healers filled an important role.

Arabic Hospitals

The hospital was a key component of the Islamic medicine. Initially inspired by the facilities in Christian monasteries, the greatest institutions were in Bagdad, Damascus, Cairo, and Cordova. The first was founded in Bagdad in 805 A.D. and was followed by others. By the 12th century 34 hospitals graced nearly every large Islamic town in their wide territories from Spain to India.

The development of hospitals changed the ministration to the poor, mentally ill, and the sick. The Islamic people understood the importance of compassionate care for the community at large. Most wealthy individuals would not have gone to a hospital but rather received care in the privacy of their homes.

In addition to successful hospital and clinic sites, the Islamic medical providers developed a sophisticated knowledge of various drugs. The Islamic pharmacological practitioners discovered and catalogued thousands of new medications. The words drug and alcohol are derived from Arabic. The first ever pharmacy was opened in Arab Croatia in 1317 A.D. and is still operational today.

Decline of Arabic Medicine

By the 12th century European physicians understood that they were being outdistanced by the Arabic practitioners. So they began to seek out Arabic texts that proffered the classic Greek Medicine. They translated them into Latin. In this role Arabs acted as intermediaries in bringing the Greek medical lore to the Western world. This became the enduring legacy of the Arabic doctors in Europe. It is no coincidence that Salerno, the first famous medical center of the Middle Ages, was close to Arab Sicily.

With the fall of Cordova, Spain to the Christians in 1236 and subsequently Bagdad to the Mongols in 1258, the Arab civilization began to decline. By the end of the 15th century, the Islamic world had become fragmented. This led to a plunge in medical and scientific progress. Nevertheless, the Muslim medical system continued to sporadically flourish until the 19th century when it receded as the tide of modern

Western medicine took hold. Some aspects of classic Islamic medical practices survive today in parts of India and Pakistan as Yunani medicine.

While the Islamic medicine rose and then declined, its influence reenergizing the canonical Greek medical theories and treatments into the West. Islamic scholars kept both the art and science of medicine alive and moving forward in the medieval period. It helped bring about a mini rebirth of healthcare after the monasteries were closed to medicine by the church. This led to the establishment of universities and a more scholastic approach to medical training throughout Europe. This academic teaching was not grounded in new theories, but based upon the reemergence of the old writings of Galen.

CHAPTER NINE
Medieval Physicians'
Training and Practices

The training of physicians underwent radical modifications in Europe during the 1000 years of the Middle Ages. Early on anyone could set themselves up as a medical practitioner and many did. As previously discussed religion and medicine were intricately entwined at this time. Priests, monks, and nuns practiced their cures in monasteries, churches and convents. This began to change towards the end of the first millennium A.D. with the church's prohibition of medicine by monks and as the achievements of Islam's golden age gradually spread through Europe.

It took another three hundred years to develop medical instruction to achieve the intellectual primacy of the university-trained physician within the hierarchy of all practitioners. While these universities enrolled and graduated only a minority of all doctors, academic curricula became systematized knowledge. This reinforced the authority of a body of medical books, concepts, and techniques that provided the basis for medical practices and beliefs broadly throughout society.

Salerno

The first generally recognized European school of medicine was in Salerno located in southwest Italy. It began sometime in the 10th century. Legend has it that it was founded by four scholars: a Latin teacher, a Jewish doctor, an Arab scholar, and a Greek. They brought to the West the teachings of Hippocrates and Galen. Salerno became the gateway for medical knowledge entering Europe. The story of Salerno illustrates how the old is also the new.

There is scarcely a segment of modern medical education that cannot be traced back very clearly to Salerno even though it began its existence 1000 years ago and ceased to attract much attention over 500 years later. One of the most enduring traditions is that Salerno was responsible for creating the title of "doctor" for medical practitioners who underwent formal university instruction.

To qualify as a Doctor of Medicine at Salerno, students underwent a highly organized training course passing from one level to another upon achieving the required level of competence. There were 3 years of initial study much like today's pre-medical under graduate education in college. This was followed by four years of medical tutelage that included instruction with specialized physicians, surgeons, medical herbalists and others. After this formal academic education, students completed a period of supervised practice mentorship similar to a modern internship or residency. This was followed by an examination that was conducted by other physicians— likely equivalent to present day specialty board exams.

At Salerno, the institution was ahead of its time in the role of female practitioners. There was a department of women's diseases. Women were admitted to the school and allowed to practice not just in gynecology and obstetrics, but also for general medical care of both sexes.

A physician's appropriate comportment and behavior was taught. One recommendation on bedside manner said, "When the doctor enters the dwelling of his patient, he should not appear haughty nor greedy, but should greet with a kindly, modest demeanor. He then should put the patient at ease before the exam and the pulse should be felt deliberately and carefully." Another suggestion advocated, "If invited for a sit-down dinner, a visiting physician was not to gorge himself, no matter how good the food and to at least occasionally inquire about the sick person. Looking desirously at a man's wife, daughter or handmaid was of course forbidden."

Other Medical Schools

Salerno became the archetype for other medical schools wishing to attract the most able physicians, papal approval and royal patronage. One of these was Montpellier on France's Mediterranean coast. Another school was founded at the University of Bologna in northern Italy in around 1200. It was in Bologna that the controversial practice of dissection first appeared on the curriculum in the later part of the Middle Ages. From there it spread to other universities spawning alterations in the understanding of human anatomy during the Renaissance.

As Bologna University matured, old time professors compromised its promise of freedom to learn. As a result, many students left and moved to Padua. Here a university was founded whose faculty was more progressive and free thinking in anatomy and surgery. This institution would emerge as the leader of medical innovations in the coming centuries.

Gradually during the succeeding 14th and 15th centuries additional medical schools were established throughout Europe. These included the University of Paris in 1100 and Oxford in 1167. They all insisted upon high levels of scholarship, unblemished morality, and taught a wide range of studies. They set scholastic standards that dominated the future of medical education in Europe. Because Latin was the universal language of the cultured world, students from different countries could be found at each of them. In the same time period books on medicine and surgery became available in vernacular languages, not just in Latin. This became a significant component in the training of lay practitioners outside the academic setting.

Impact of Formally Trained Physicians

In spite of ample opportunities, there were only a small number of students that completed the course of instruction from the university medical schools. At Bologna in the early 15th century it took 7-10 years of study to qualify as a physician. An average of only four doctors each year graduated. The considerable amount of time to achieve a degree discouraged many who were not from wealthy families to pursue a medical career. An interesting picture of the small number of formally trained within the population is reflected in the ratio of physicians to townspeople. In Florence, Italy in 1338 there were only 60 licensed physicians for a population of 120,000.

New physician graduates were subject to legal regulations. Laws were enacted in order to prevent unfit and unworthy individuals from practicing medicine solely for their own personal benefit or to the detriment of their patients. Other statutes regulated medical fees, required free attendance on the poor, and promulgated rules for pharmacies.

Surgeons and Their Craft

Medieval medicine was at its nadir in the field of surgery. Establishment medical doctors viewed the practice of surgery with disdain. It was considered manual labor, more compatible with the tasks of a tradesman than a physician. Surgeons early on were not educated as medical professionals. This pejorative thinking about surgery was enhanced by church doctrine. It prohibited the shedding of blood and opposed dissection to learn human anatomy. This axiom prevented the natural evolution of surgery as a science.

The prevalence of this negative attitude meant medieval surgeons initially came to the profession without any formal medical study. They emerged from the ranks of the town barber, bath-keeper, hangmen and other common jobs of every description. From these groups the barber-surgeon class developed. They often traveled from one community to another providing their services. The itinerant nature of their practices likely came about so that there was amble opportunity for the self-proclaimed surgeon to quickly slip out of town if their patient's condition deteriorated or died.

When people sought the help of a barber-surgeon, they were generally in need of immediate care. The time required for surgical intervention had to be brief since there was no adequate system for deadening pain. Relief was limited to alcohol or some kind of sleep inducing mixture of plants mixed with wine. Patients sometimes solved the pain issue themselves by passing out from the agony of their condition. In this instance, working on someone who was unconscious made the surgeon's job easier.

Nearly all common procedures were carried out. The barber-surgeons stitched cuts, pulled teeth, dressed wounds, applied ointments, lanced boils, stopped hemorrhages, gave enemas, repaired hernias, removed nasal polyps, excised hemorrhoids, and cut out tumors. In addition, they performed bloodletting, cupping and cautery. Some barber-surgeons became reasonably competent in these various tasks because they frequently accompanied armies to war. Under the heat of battle, they became experts in repairing wounds, performing amputations and other everyday surgical interventions.

By the 13th century barber-surgeons organized themselves into professional guilds and implemented apprenticeships to study their craft outside the purview of university medicine. Most were men, but women trained as well.

Surgery in Medical Schools

Barber-surgeons chiefly confined their practices to the northern part of Europe. In this geographic area their craft was a completely separate division of medicine from the more formally trained physicians. This arrangement and the barber-surgeons' presence in Italy and southern France were less evident. The traditional medical teaching persisted in these countries. Surgery was included as part of the overall scholastic curriculum at most of the university medical schools. But early on, because of the perceived lack of prestige, very few university-trained doctors actually practiced surgery to the level of the unsophisticated barber-surgeons.

Not withstanding the barber-surgeons' untrained presence, it would be a mistake to conclude that there was no advancement in surgery during the Middle Ages. While this was mainly true, new historic insight has been gained in the recent republication of old texts and the restudy of medieval documents. This new assessment makes it clear that surgery was successfully cultivated to some degree during the late Middle Ages. This occurred with the continued development of the university medical schools. Some students began seriously studying the craft and a new breed of professors contributed to surgical knowledge.

In this pedantic setting the relationship between non-surgical university physicians and eminent surgeons was contentious. Middle Age academic surgeons laid claim to both surgical skills and to the usual scholarly learning of all physicians. Because they possessed "a good eye, a steady hand, and a sharp blade" many felt superior to their non-interventional colleagues. As time passed in southern Europe the gulf between university trained surgeons and physicians lessened. It became apparent that surgery was a desirable expertise for a physician to acquire. Elsewhere the gap between the two classes of doctors widened.

Beyond Italy, surgery was excluded from the academic curriculum. In northern Europe surgical training and practice were organized on a guild basis. This left most surgical procedures in the hands of the barber-surgeons, not with the academically trained. Eventually in locations like Paris and London the conflict between university surgeons and physicians receded in a united dislike towards the untrained barber surgeons. This division between medical and surgical practitioners stopped after many centuries of conflict.

The political aspects of the surgical practice within and beyond the universities plus new books on the craft had limited
bearing on the actual performance of successful surgical
procedures for the populace. In the Middle Ages and to a
certain extent even today positive surgical outcomes depended on manual adeptness, good judgment, practice and a
degree of luck.

Innovative Medieval Surgeons
Henry de Mondeville and Guy de Chaulic

Frenchman Henry de Mondeville (1260-1320) said, "it is
impossible to be a good surgeon if one is not familiar with
the foundation and general rules of medicine; and it is
impossible for anyone to be a good general physician who is
absolutely ignorant of the art of surgery." De Mondeville was
a military surgeon who advocated simple bathing of wounds
and immediate closure, followed by dry dressings with minimal loss of flesh or skin. This new approach met with strong
opposition from those who believed the old Greek method
that taught "laudable pus" was beneficial to the healing process. Mandeville's overture demonstrated that some doctors
were prepared to think for themselves and even challenge the
words of Galen and other ancient writers. This was the beginning of important changes in medicine.

A disciple of de Mondeville was Guy de Chauliac (1298-1368).
He was considered the most prominent surgeon of his
generation and is now known as the "father of surgery." His

great work, the *Chirurgia magna*, was an astounding exercise in surgical erudition. It covered anatomy, inflammation, wounds, ulcers, fractures, dislocations and miscellaneous diseases belonging to surgery. It was translated into several languages.

De Chauliac exhibited the practical qualities of being a surgeon. He devised a bed frame that made it easier for patients to turn. He recommended new ways to splint a fracture, and he used weights to stretch a broken limb to help it heal more cleanly.

Beyond surgery, his writings also contained fascinating first-hand details about the plague and a potpourri of treatments outside of his specialty. In his discussion about impotence de Chauliac said that "if physical causes could be ruled out, he would hire an "experienced" woman to shower the troubled couple with wine, massage and talk to them suggestively for three days."

Middle Ages Hospitals

Some medical historians consider that one of the few noteworthy medical accomplishments of the Middle Ages was the hospital. Although institutions like hospitals had existed under the Romans and in the Byzantine Empire, they cannot be compared in magnitude and importance to those in the Christian hospitals that started in the mid-portion of the medieval era.

Early on small healthcare facilities had religious origins. They were within the monasteries and convents. These installations served as infirmaries for the monks, nuns, surrounding populace as well as lodging places for travelers. Their late medieval successors were more impressive and continued in the overwhelming majority of instances to have a religious affiliation. Some hospitals that began as small additions to monasteries became major institutions like St. Bartholomew in London.

The great age of hospital fabrication coincided with the flourishing new universities. By 1400 in Florence, Italy alone, a city of 130,000 inhabitants, there were thirty separate medical institutions. The hospital movement spread throughout Europe. Every city in Germany with a population of more than 5,000 people had at least one hospital.

The size of these different facilities varied. Some could only accommodate 10 to 50 people, while other like St. Leonard's in York, England could care for 225 individuals. Hospitals were set up as large dormitories. Men were separated from women, but as was common practice in many households, more than one person was frequently assigned to a single hospital bed. These healthcare facilities were primarily for the poor. When sick, the wealthy would have stayed at home.

The majority of these early hospitals were not principally medical facilities, but rather they were philanthropic institutions offering "hospitality" and refuge for the old, the disabled and homeless pilgrims. The slow transformation from charitable hospital into a real medical institution had a meager beginning in Italy in the waning years of the Middle Ages. The process started when the administration of hospitals was gradually taken out of the hands of the religious orders by local city governments. The medieval versions of hospitals never fully became healthcare facilities in the modern sense. This did not occur until the late 19th and early 20th century.

Uroscopy, Astrology, and Alchemy

Medicine that was taught and practiced at this time was primarily based upon Galen's theories and teachings. Diagnosis was made, and then to rebalance the four humors remedies were primarily diets, drugs, bloodletting, and purging.

One of the crucial procedures used in the diagnostic process was uroscopy. In addressing the merit and importance of the examination of urine, a medieval writer said, "Through science I can show you the reasons of the whole universe." This seems a little farfetched particularly since doctors then had no understanding of the chemistry behind the value they placed in examining the patient's urine. It nevertheless became such an integral part of medicine at the time that the glass beakers used by physicians to hold and examine their patients' urine became a symbol of the medical profession. It was believed that urine was a "liquid window" that could unlock the body's secrets and was the best way to diagnose disease.

Physicians considered the urine's color. There were 20 recognized hues and a rainbow like standardized chart to compare

with a specimen. The urine's smell and texture was important—
was it thin and watery or thick and greasy. For example, red
colored urine would mean blood, and grit that was visible may
signify kidney stones. With the passage of time, the analysis of
urine assumed even more importance. A French physician
wrote a 347-line poem titled "On Urine" that all medical
students had to memorize.

For reasons unknown, the switching of urine samples became
a big problem and a scandal. Few deceivers, however, could
mislead Arnold of Villanova, a physician, astrologer and
alchemist. He said, "To the individual who brings the urine
… keep your eyes straight on him or his face … if he wishes
to deceive you …. the color of his face will change, and then
you must curse him forever and in all eternity."

Urine was prophylactic for some—it was a health drink. Pope
John the 21st, the only medical doctor to become a pontiff in
1276, drank it religiously until he died when the ceiling he
designed himself fell on his head killing him.

Beyond examining a patient's urine, the use of astrology
became deeply ingrained in a doctor's practice. This was the
concept that the divination of the stars and planets could
predict human affairs by their various positions and aspects.
By the late Middle Ages in both theory and practice astrology
was part of the syllabus of Europe's leading medical schools.
Laws were passed requiring that doctors carry the latest
charts and horoscopes in their medical bags.

In contrast to a doctor's own hit or miss observations, astrology was thought to be able to predict with near perfect accuracy the exact time when a patient should be treated or have a procedure carried out. Migraines, for example, were best cured on April 3rd, but blindness on April 11th.

Many surgeons made sure that the time was right before doing a procedure. For instance, the exact phase of the moon had to be considered for brain surgery. It was believed that the body's four humors are agitated by the different phases of the moon. As a result, it was thought that the brain within the skull waxed and waned as the waters in the oceans do, so the right time to operate was important. If one were to have brain surgery in the Middle Ages, the patient would want to be certain to undergo the procedure during the appropriate moon phase to survive the very risky operation.

Alchemy did not have as many advocates as astrology and uroscopy. It was a curious mix of science and magic. It had assorted lofty aims that ranged from changing ordinary metals into gold to curing all illnesses. The objectives of the alchemist varied, but the underlying thread was always change or transmutation for the sake of improvement. This usually meant one of three pursuits:

(1) physically altering a common substance into a precious one.

(2) spiritually bringing light to darkness.

(3) medically giving good health to the sick.

There was a tendency toward esotericism among alchemists. They restricted knowledge to a few privileged practitioners who could mystify ordinary people. Yet alchemists helped develop real world skills. They extracted ingredients from plants, animals, and rocks. They mixed, boiled, and purified elements. These and other procedures are still practiced today.

Among the alchemists' medical aspirations were to find a universal panacea to cure all ills and at the same time develop an elixir to maintain youth. By the 1700s faced with the vigorous application of the scientific method and the success of the young subject of chemistry, alchemy faded into an occult pursuit.

CHAPTER TEN
Middle Ages Diseases
and Treatments

L ife was tough for all except the rich in the Middle Ages. In 14th century Europe there were nearly 80 million people—about 11% of today's population. Food was scarce and expensive, wages low, and living conditions and the environment provided an ideal breeding ground for disease. Most did not enjoy long lives. Infection and accidents killed nearly half of all children before the age of five. If one made it into adulthood, the majority died in their twenties and thirties. The causes of death included women in childbirth, men in war, accidents and everyone else from fatal illnesses like leprosy, malaria, dysentery, small pox and tuberculosis.

There was also influenza, intestinal worms and scabies—an itchy, disfiguring skin infection spread by mites. Even simple colds could be killers if a pulmonary infection resulted. About 25% of the population lived to over fifty years of age. These individuals were apparently endowed with a robust internal immunological system.

Treatments

Treating those that became sick in the Middle Ages was a mixed bag. The university-trained doctors' approach was similar to their ancient Greek and Roman counterparts. They suggested dietary changes, plenty of exercise, and as previously noted bleeding and purging to restore the proper balance of the patients' humors.

The most common remedies were based on minerals, animal parts, plants or herbs. Physicians and every wife and mother were expected to have detailed knowledge of such cures. Monasteries, convents and most homes had herb gardens for growing plants that could be used in medicines. Honey was a common ingredient used to kill infection since the time of ancient Egypt.

One medical compound that was universally used during the Middle Ages as well as into the 19th century was Theriac. It was an old medication that was initially concocted in the first century A.D. as a general antidote for poisons. Over its long history it was prescribed for almost every known aliment from the common cold to cancer. Its efficacy is questionable, but it worked for many because of the placebo effect and the mild pain reduction generated from the opium that was usually part of its formula.

Leprosy

Some untreatable diseases loomed large both in reality and in the public's imagination as well. Such was the case with leprosy. Its physical symptoms—scaly flesh, mutilated toes and fingers, boney degeneration and unclean appearance—led to deeply punitive attitudes and strict laws in medieval Europe.

Those afflicted were forbidden all normal social contacts and became targets of shocking rites of exclusion. They could not get married, were forced to dress distinctively and had to sound a bell warning others of their presence. People with the disorder were segregated into special houses and hospitals called leprosaria. Today, we know leprosy is caused by a bacterium and is curable.

Bubonic Plague or Black Death

There were no ancient diseases that compare to the devastation caused by the Bubonic Plague or the Black Death that occurred from 1347 to 1353. During these six years between 40-50 million people died—nearly 50% of the population. The pandemic was carried by merchants traveling busy trade routes from Constantinople (Istanbul, Turkey) and ports of Africa. As ships reached their destinations in Europe it spread throughout the continent. Only Iceland and northern districts of Scandinavia escaped. On one trip alone, it was reported that out of the 332 people making the journey, just 45 were alive when the boat arrived from the near East.

Medieval people did not understand what caused the pestilence. They choose to believe that it was God's punishment for their sins. Doctors could not fabricate any better theory. They evoked astrology that said the plague was caused by dangerous planets—Mars, Jupiter, and Saturn—passing through the constellation Aquarius on March 20, 1345. Other doctors postulated that garbage and dead animals left on the city streets created a 'miasma' or poisonous air that initiated the devastation.

It was not until the late 19th century that the bacterium *Versina pestis* was identified as the cause of Black Death. Epidemiologists believe that infected rodents carried the bacteria. Fleas picked up the microbe living on rats and transferred it to humans via fleabites. A flea could discharge up to 24,000 bacteria in one bite. During the epidemic the illness infected all in its path that were susceptible. It eventually burned out after reducing the population to those who formed some immunity.

While too late for those in the Middle Ages, an antiserum was developed in 1896 that was successful in about half the cases. The introduction of the antibiotic streptomycin in the 1940s increased the cure rate to about ninety-five percent. There are still sporadic instances of the plague. In 2013 a boy in Russia died after eating a plague-infected marmot. In 2015 in the United States there were fifteen cases of plague infection including four deaths.

The Plague—A Complex Disorder

The plague of the 14th century was not a simple disease. It presented in three forms. The most common variation started with swelling or buboes in inflamed lymph nodes of the arm-pit or groin. They varied from small to as large as an apple. These lumps spread over the entire body and often turned black and purple from internal bleeding. Their presence was accompanied by high fever, headache and fatigue. Most in-fected individuals died within a week. The appearance of buboes came to be known as a certain sign of death.

The second variation involved the lungs. It was much more virulent. This form of the contagion traveled from person to person through the air—not by a fleabite—attacking the respiratory system. It caused death in one or two days. The most infectious third type was a septicemia –an infection within blood. It was extremely lethal causing death very quickly.

The doctors were mystified how to treat it. There was noth-ing in Galen's teachings to help. Ineffective treatments were tried including avoiding foods that were hard to digest and purifying the air with roses and cloves to reduce the noxious vapors. In some countries officials attempted to be proactive by requiring ships arriving from infected ports to sit at anchor for 40 days before landing. The name for this practice—quarantine—was derived from the Italian words *quaranta giorni* meaning 40 days.

Beak Doctors

During the plague years, many if not most doctors hid from their patients or simply fled. Those dedicated brave clinicians that stayed behind tried to safeguard themselves from the contagion. They wore elaborate costumes that they hoped would reduce their exposure to the miasma that they believed caused the illness. Their protective clothing included birdlike masks that featured beaks up to a foot long that functioned like a crude gas mask. Eyeholes in the mask were covered with red-tinted glass to stare down the evil eye.

A poem about these "beak doctors" read:

> *When to their patients they were called*
> *In places by the plague appalled .*
> *Their hats and cloaks of fashion new*
> *Are made of oilcloth, dark of hue,*
> *Their caps with glasses are designed*
> *Their bills with antidotes all lined,*
> *That foulsome air may do no harm*
> *Nor cause the doctor man alarm.*

Consequences of the Plague

The social and economic effects of the plague were monumental. When the main epidemic ended in 1353, the shocked and anxious survivors faced an uncertain future. Their families and communities had been shattered. The tragedy did not immediate cease. There were repeated out-breaks, but to a lesser degree, over the next two decades and beyond. England experienced its final outbreak in 1665 and France in 1720. The 14th century Black Death provoked realignments within medieval society. That process had started long before the plague broke out. It just became more apparent in the wake of the epidemic's destruction.

As the population declined, farm laborers became scarce and land became vacant. This allowed peasants to demand higher wages. The balance of power shifted. The Lords and rich landowners were faced with considerable adjustments. They had increasing difficulty enforcing old claims made on serf families and were confronted with peasant rebellions. Farming was less profitable and the nobles sought new capitalist enterprises. More broadly the people began to assume a new level of popular participation in local and national government. While this was an encouraging development, it came about at a very big price.

Mixed Conclusions about Medicine in the Middle Ages

Was the medieval time just an epoch of obscure inconsequential accomplishments or did it have relevance in the expansion of medicine as a whole? It was at its best in the years that the golden age of Islamic Medicine flourished. The events and personalities during that time was a shining light that added some innovations and acted as a catalyst for future reforms. At its worst, it was when the church's authoritative doctrine discouraged major revamping in treatments and new discoveries. This negativity was slightly balanced with the positive role religion had on the expansion of hospitals in the latter half of the era.

The most enduring legacy of the Middle Ages was the establishment of medical education at the university level where scholars began to question old ideas. While some fundamental medical advancements were just ahead, the core practice of medicine for the populace continued to promote ineffective ancient methods and theories of Galen.

CHAPTER ELEVEN
The Early Renaissance

The Renaissance was the period in European history from 1500-1700. It was an age of new thinking in the arts and discoveries in diverse fields. The seeds of reform were planted in the late Middle Ages following the great plague that brought increased wealth in many sectors. This along with the exploration of new territories in better ships, the invention of the printing press and modifications in warfare pushed scholars in all fields to reevaluate and improve on ancient ideas and techniques.

Painters, philosophers, and poets commended the beauty of the human spirit as the revival of Greek learning and culture took place. This produced a humanistic approach to their work. Mankind was to view the universe differently.

Martin Luther's reformation in 1517 weakened the once unshakeable authority of the Roman Catholic Church. For centuries the church had taught mankind to renounce worldly goods for the sake of eternity. Renaissance man reversed that attitude and exhibited an insatiable curiosity for the materiality of the here and now. This fostered a new realism in multiple fields of study and different aspects of society.

With all these adjustments in so many areas, the groundwork was laid for the early beginnings of the scientific revolution. In medicine, by the end of the Renaissance era the work of doctors and scientists overturned many of the traditional medical beliefs. This established a foundation of facts that became the launching pad for the astounding jump in healthcare knowledge that followed in the 19th and 20th centuries.

At the onset of this discussion of Renaissance medicine, it would be a mistake not recognize a significant contradiction that occurred during this time. In spite of the leaps in scientific medical knowledge that occurred, it had little influence on ordinary doctors and their healing of the populace. With some obvious exceptions mainstream medicine still followed the old ways and as one historian noted, "remaining ignorant and pompous as ever."

Leonardo da Vinci

Art was one of the initial fields in which the new realism became dominant and effected medicine as new trends in the study of human anatomy emerged. The Renaissance was a time when science and art were not considered polar opposites. In fact, the relationship between artists and medical people was so close that doctors, apothecaries, and painters in Florence, the initial epicenter of the Renaissance, belonged to the same guild. It is difficult to describe this interrelationship in detail, but the career of Leonardo da Vinci (1452-1519) illustrates it admirably.

While he is best remembered for his paintings, Leonardo contributed to many fields including architecture, engineering, military weaponry, human aviation, and botany. Unknown in his lifetime were his contributions to the field of medicine. Beautiful and anatomically accurate drawings of various parts of the human body filled many of Leonardo's notebooks.

While the church still prohibited human autopsies on a regular basis, a Veronese anatomist, Marcantonio della Torre gained a special ecclesiastical sanction to perform post-mortem examinations. He asked da Vinci to work alongside him to prepare illustrations for a text on anatomy. When Della Torre died unexpectedly, Leonardo assumed both tasks, performing the dissection and doing the illustrations. Because he personally had not gained permission for necropsies from the church, he worked in secrecy in the cellar of the mortuary in Florence dissecting and drawing as many as 30 human bodies.

He made numerous sketches of the human skeleton, muscles, heart, reproductive system and other internal organs. He produced one of the first scientific drawings of a fetus in utero. Leonardo's dedication to observing and realistically recording individual body parts as they performed mechanical activities was the feature that made his work so exceptional.

In 1651—nearly 150 years after his death—many of Leonardo's anatomical drawings were published for the first time. Unfortunately, his work had little direct impact on the advances in anatomy that occurred in the 16th century. His artistic renditions of the human body that have survived forged the basic principle of modern scientific illustrations.

The Revolution in Anatomy

Before the remarkable strides in human anatomy could be made, there needed to be a radical change in attitude about dissections. The devastation of the plague at the end of the Middle Ages helped alter the church and medieval societies' belief that it was a sacrilegious to interfere with the mortal remains of man.

Initially, the church permitted autopsies only on plague victims to determine the cause of death. In 1537 the laws were altered to permit postmortem dissections on an as needed basis. This led to the inclusion of anatomy as a part of medical schools' curriculum. It was difficult to obtain cadavers since the church regulated the number of bodies that could be made available. This led to innovative and often illegal ways to obtain corpses for examination. Medical students raided graveyards and the gallows to procure dead bodies for their studies.

Andreas Vesalius

Into this turbulent sea of change in 1543 came a twenty-eight-year-old author, Andreas Vesalius (1515-1564). His book, *De humani corporis fabrica* (On the Fabric of the Human Body), revolutionized human anatomy and medicine. It stood as a monument for metamorphosis in the history of science as a whole. Based on his own dissections and physiological experiments, his findings stirred a storm of controversy and sent medicine on a new path.

Vesalius' book gave mankind its first accurate description of the structure and workings of the human body. It exemplified a revitalized scientific method that allowed others to correct or add to what he had found. It demolished Galen's anatomy ideas that European and Arabic physicians had relied on as infallible. The great Galen, Vesalius wrote, "never dissected a body of a man who recently died."

Vesalius was born into a family of physicians in Brussels, Belgium. He took an early interest in how living things worked. While still a boy, he undertook dissections of small animals on his mother's kitchen table. At the University of Paris Medical School, he studied under the well-respected Jacob Sylvius (1478-1555). Sylvius like many other contemporaries was absolutely devoted to Galen's teachings. Sylvius ignored any discrepancies between the old master and the actual autopsies that he was undertaking. Vesalius on the other hand detected the differences and began to speak

openly about his disagreement with Galen's anatomical theories. Vesalius told fellow students that they could "learn more
at a butcher shop" than from professor Sylvius.

Not unexpectedly, Vesalius moved to the University of Padua
early in 1537. He took Padua by storm earning his doctorate in
medicine with the highest possible honors before the end of
that year. A day after graduating, he accepted the university's
offer of a position and begun to teach anatomy and surgery.

From his first class, the 22-year-old professor broke with
tradition by introducing a new lecture technique. His anatomy
lessons spurned the customary lectern and the usual assistant
who carried out the autopsy while the professor spoke not
touching the corpse. Vesalius lectured directly to the students
while he personally performed the dissection pointing to the
various structures that he was discussing. His fame spread
rapidly, and he was invited to give lecture demonstrations at
other medical centers.

Vesalius' series of anatomical lessons at Bologna's medical
school in 1540 proved to be his epiphany. While trained as a
devoted Galenist, his observations during the multiple dissections of human bodies did not match with Galen's time-
honored teachings. With further study, Vesalius began to see
a pattern in Galen's many errors.

Time after time Vesalius found that the ancient anatomist's
written descriptions of human anatomy did not jibe with his
findings, but Galen's observations did match apes, dogs, or

sheep. He realized that Galen had incorrectly assumed that his work on these animals held true for the anatomy of the human body as well. Vesalius' controversial evaluation excited the students, but it shocked the older faculty members who as one medical historian remarked, "had turned their backs on the future."

For the next several years he worked feverishly on what he knew would depict anatomy correctly for the first time. Vesalius reached out to the master painter Titian who provided one or more of his ablest students to observe his dissections. They captured the details in vivid lifelike poses that brilliantly illustrated the human skeleton, muscles, tendons, veins, arteries, nervous system, abdominal and reproductive organs, the heart, lungs and the finally the brain.

De humani corporis fabrica libri septum

The publication in 1543 *De humani corporis fabrica libri septum* (On the Fabric of the Human Body in Seven Books) marked an irreversible turning point in medicine. For the first time the understanding of health and treatment of disease could be rooted in an accurate knowledge of human anatomy. The 663-page book itself was a masterpiece that was divided into "seven books" (equivalent to chapters) that summarized Vesalius' extensive findings. The manuscript's pages were large (16.5" x 11") as were the 400 plus individual images. The quality and harmony of the art and typography make it one of the gems in the history of printing.

Among his many discoveries were that men and women have the same number of ribs, the mandible is a single bone not two, the nerves are solid not hollow and end or originate in the brain. His revolutionary work only went so far—there were errors. Vesalius' dissection gave him a complete understanding of anatomy, but many mysteries remained on how the body functioned. Vesalius failed to divorce himself from Galen's model about blood flow and the vascular system. He misunderstood that the purpose of breathing was not to cool the blood, and that the digestive process did not involve some method of "cooking" the food to digest it.

Much to Vesalius' chagrin Galenic anatomists led by his one-time mentor Jacob Sylvius initially spurned his original scientifically proven study. Some religious leaders were even more strident in their objections. Sylvius spearheaded the effort of medicine's old guard to discredit Vesalius. Their argument was simple "since Galen could not be wrong, Vesalius must be."

It would be an enormous understatement to say that Vesalius did not react well to criticism. He responded with an act that seems to have been a rage-driven symbolic suicide. In the last days of 1543, he piled up his anatomical studies, his unpublished commentary on Galen, his preparatory notes for future works, and burned them. The brilliantly driven young medical revolutionary symbolically disappeared in that dramatic bonfire. From the ashes emerged a new, almost unrecognizable conservative Vesalius. He left Padua forever, married, and donned the robes of a respectable physician in the court of

Charles V and Philip II of Spain. During his years in royal service, Vesalius became Europe's most respected physician and grew wealthy.

By the end of his life Vesalius was disillusioned with practicing in Spain's royal court. He negotiated a return to Padua to reclaim his position as professor of anatomy. Going back by ship from Spain he fell ill during a terrible storm. He was put ashore on the Greek Island of Zane, where he died alone and was buried in an unmarked grave. Fate did not give him a second chance.

Vesalius' legacy was that he established the modern science of anatomy. He corrected long-held misbeliefs, introduced new discoveries and inspired a fresh breed of physicians. There may be no more fitting memorial for Vesalius that the words that appear on the side of a tomb in one of the famous illustrations of the human skeleton in his book. It says, "Genius lives on—All else is mortal."

Other Anatomists:
Falloppio, Eustachio, and Wirsung

After the publication of Vesalius' book, practical anatomy advanced on a broad front. His pupil and successor Gabriele Falloppio (1523-1562) added new observations and corrected errors in both Galenic and his mentor's anatomy. Falloppio is best known for his remarkable descriptions of the female reproductive system—the fallopian tubes are his namesake.

He also studied the internal ear and in particular the semicircular canals that are responsible for maintaining the body's equilibrium.

Another pioneer anatomist was Italian Bartolomeo Eustachio (1520-1574). His place in history may well have been in the same league as Vesalius if his works had not been misplaced. Eustachio's engravings and comments were discovered 150 years after his death. They were on an equal level of scientific importance as Vesalius in descriptions of the base of the brain, structure of the larynx and inner ear of which a part is named after him.

In 1642 German Johann Wirsung established one of the consequential milestones in anatomy and medicine with his discovery of the pancreatic duct that received his appellation. His story could also have been a script for a murder mystery plot on television.

But first to the science—the pancreas had always been an enigma for early physicians. Although Wirsung was unsure of the purpose of the duct that he found, his assessment was significant. He established the role that the pancreas had as a gland. It secretes a fluid that aids in food digestion. Wirsung made the mistake of not publishing his findings. Instead, he sent illustrations of his findings to other European anatomists for their comments.

The story does not end here. A year after his discovery, Wirsung was shot and killed while chatting with a neighbor near his home. A jealous mentor was accused of the crime shortly after the murder, but he was acquitted and set free. Historians subsequently conjured up many conflicting theories on the identity of the assassin, but the slayer was never found.

CHAPTER TWELVE
Renaissance Diseases
and Practitioners

P eople did not live long in the Renaissance period. The number of childhood deaths was very high as babies and children were extremely vulnerable to infections. There was also a significant mortality rate among women in childbirth. Overall less than half of the population of Europe lived to the age of forty.

While common maladies like dysentery, colds, and infections remained, there was also an evolving pattern of illness as Europeans began making long voyages. In the process isolated people were suddenly immersed in a broadening germ pool and new diseases emerged. Old disorders like scurvy, smallpox and syphilis became more virulent.

Scurvy

Scurvy was a disease that had been around for a long time. It was described by Hippocrates and appeared among those who joined the Crusades. It presented with spots on the skin and bleeding from the mucous membranes particularly around the mouth. As it progressed the victim's muscles become rubbery, making it arduous for the afflicted individual to move.

In the Renaissance, scurvy's prominence increased because of the expansion of sea travel. Today, it is known that scurvy results from a deficiency of Vitamin C that is contained in fruits and vegetables. In the 16th century voyages of two years or more, ships could not carry enough fresh food on board . So scurvy abounded.

There were isolated instances where explorers stumbled on a cure for the disease. One was that of Jacques Cartier who found the St. Lawrence River. Because of the river's ice in the winter of 1536, he made camp near an Iroquois village. His crew exhibited symptoms of the illness that now is known to have been scurvy. Cartier approached the Indians who did not seem to be afflicted. After some persuasion and mistrust the Indians gave him their secret medicine—a tea made from the leaves of the white cedar tree. Cartier's men who drank the tea felt better. The remedy has been analyzed since then and has been shown to contain a high level of Vitamin C.

In the 17th century physicians began to understand that scurvy resulted from an unknown dietary deficiency. It was not until after the Renaissance in 1754 that a definitive link was established. A Scottish navel surgeon, James Lind, in the world's first controlled clinical trial on board the HMS Salisbury, proved the effectiveness of citrus fruit to control and prevent the disease.

Scurvy has not completely disappeared from the modern world. It can occur when there is poor nutrition. A contemporary practicing physician recently reported about a patient who developed scurvy. Her diet was solely Twinkies, nothing else.

Smallpox

Of all the human diseases, smallpox has perhaps the most claims to fame—or rather infamy. It has been featured in all of recorded history, killed billions, and inflicted lasting suffering on billions more. It was the first infection to be immunized, as well as the first and currently the only major global disease to have been eradicated.

In 16th century Europe smallpox was common, relatively mild and rarely fatal. In many communities it was considered to be one of the various illnesses suffered during childhood. In the 17th century, smallpox became more virulent likely from a new strain that came from Asia. Various areas, particularly Italy and England, were hard hit over a 20-year period. The disease followed its own destructive course. Only a few recovered, most died. Those that survived were maimed or blinded by the illness. Renaissance physicians did not understand its cause nor how to treat the disorder. But they were aware that those who survived the infection obtained immunity.

Scientist now recognize that the virus Variola causes small-pox. It spreads through inhalation of airborne droplets and direct contact with bodily fluids from infected individuals. Its name "small pocks" refers to the skin eruptions or sacs that leave pitted pockmarks. This term was introduced in 15th century England to distinguish it from syphilis that was called the "great pocks."

The effect of smallpox in Europe was significant. But it paled to the devastation that occurred in the new lands where explorers landed in their ships. The native populations had little or no immunity. After the Spanish conquerors arrived in Central America in the 16th century, almost half the Aztec population perished from smallpox. The Inca population of Peru fell from 7 million to 1.5 million.

With no cure known, the Renaissance doctors looked for methods that might prevent the disease. In 1522 Paracelsus, a brilliant but controversial doctor of the age, traveled to Turkey where he learned of a peasant remedy to prevent smallpox. While Paracelsus' method worked it did not become accepted because he also used similar inoculation techniques with other illnesses without success.

Inoculation was introduced in Europe and then America after the Renaissance in the 18th century. This was followed by the more effective vaccination discovery at the end of the same century.

Syphilis

In 1495 an unfamiliar disease swept through Europe that was transmitted sexually or from mother to child in the uterus. Causing painful sores, madness and death, syphilis exacted a terrible toll for over 4 centuries.

There are a number of theories on how syphilis traveled to Europe. Those with an astrological bent suggested that it was caused by a misalignment of the planets. Others postulated it was a combination of leprosy and gonorrhea that were both known at the time. Another speculation holds that the African slaves brought syphilis to Portugal in 1442.

The most popular hypothesis is that sailors conveyed it back to Europe from Columbus' first voyage to America in 1492. This is supported today by a number of independent studies that found the presence of syphilis is ancient American skeletons. But the most compelling evidence is a 2008 genetic study that demonstrated the syphilis that spread in Europe was linked to a strain of yaws—a similar presenting disease— that existed in South America.

There is an interesting observation by one modern researcher on the importance of comprehending the epidemiological and geographic origin of syphilis as a major killer during Renaissance Europe. He stated, "learning about it is not just for biology, but for understanding social and political history, and for today one of the important early examples of globalization and disease and how globalization remains a significant factor in emerging contagions."

From the 16th to the 19th century attempts to identify the cause of syphilis was hampered by its early confusion with leprosy and gonorrhea. Today, the etiology is well known. It is the thin spiral shaped organism, *Treponema pallidum* that spreads through sexual contact.

Early on mercury emerged as a cure and remained so until the 20th century. It was usually combined with other ingredients like turpentine, incense, sulfur and lard. Those with syphilis sat in a large tub where they were rubbed with mercurial ointments several times a day. Shakespeare notes the torments of syphilis and makes reference to the "tub of infamy" in his play about Henry the Fifth. The nursery rhyme "Rub-a-dub, three men in a tub" is thought to be about syphilis.

Most modern venereal disease victims are reluctant to discuss their aliments, but this was not the case during the Renaissance. The culture at the time thought nothing of sexual promiscuity among the upper classes. So there was no particular stigma associated with a sexually transmitted disease. With this open attitude about extramarital sexual activity, the illness could be dealt with more directly, but not particularly effectively.

In 1906 the invention of the Wasserman test was a break through. It was able to identify victims of syphilis even if there were no identifiable symptoms. The first non-mercurial based treatment, Salvarsan, was discovered in 1909. This was the first drug to target a specific pathogen. It remained the principal anti-syphilitic cure until 1943 when penicillin became readily available. Syphilis is controlled but not eliminated today. It continues to be a serious health problem. In 2015, it caused about 107,000 deaths worldwide.

The disease became known as syphilis because an Italian physician Girolamo Fracastoro (1478-1553) was fascinated by epidemic illnesses. In 1530, he published a fictional narrative poem that described a disease suffered by a handsome young man he named Syphilis. As Fracastoro told it, Syphilis brought about his own illness because he cursed the Sun. To punish him and other men for this blasphemy, Apollo, god of the Sun, shot deadly rays of disease at the Earth. Syphilis was the first victim, but it soon spread to others, including the king.

Within the Latin verse, Fracastoro presented the symptoms, the pattern of the infection, and the recommended treatment. Through the poem the signs of the illness became better known, and eventually the shepherd's name was adopted as the appellation for the disease.

To remedy the multitude of diseases Renaissance humanists believed it could only occur with the restoration of medicine to its Greek purity. An opposing counter-view challenged this thinking beginning with Vesalius. While his and other early anatomists' radical rhetoric exposed factual errors, they failed to address many major segments of ancient medicine. A fresh start was needed.

Paracelsus

Vesalius and his successors did not address anything substantive about Galen's humoral theory. Instead, the frontal assault on this ancient hypothesis came from a true Renaissance rebel, Philippus Aureolus Theophrastus Bombastus von Hohenheim (1493-1542) or as he called himself Paracelsus. In his early 30s he shortened his name to Paracelsus that meant "greater than Celsus." He was a famous encyclopedists of the first century A.D.

Paracelsus was one of the most contradictory figures in a contradictory age. He reflected the violent and confusing aspirations of the common man in the early 16th century. Paracelsus set out to do nothing less than smash the venerable establishment of Galenic medicine with its four humors, rote diagnosis, ritualistic bleeding and purging, and concocted prescriptions. Symbolizing his break from the past, Paracelsus was the first prominent physician to abandon Latin and use the vernacular in his medical writings. He was brilliant but controversial. He introduced fascinating new problematic theories, but his ideas were slow to be accepted because of his rogue personality.

Paracelsus' life was a paradox. It would have made a great movie. He preached self-denial as a measure of personal and spiritual discipline but would drink with the miners and team-sters (the 16th century equivalent of today's truck drivers). He berated his colleagues for their arrogance and false claims but declared himself the monarch of medicine. Paracelsus lived a vagrant's life. He roamed from Sweden to Egypt and from Russia to England—first by choice, but later because he had alienated the medical, religious, and civil authorities every-where he went.

Much of his uniqueness can be traced to his childhood in Switzerland where he came from a noble family in decline that beat to a different drummer in a time that was steeped in tradition. His illegitimate father ministered to the workers of the mines even though he had no formal medical credentials. His grandfather was a commander during the crusades. He was known to have had a fierce temper and quarreled with the wrong people. His grandson inherited these traits.

Just where and how Paracelsus acquired his formal education remains unclear. What is known is that his formal schooling was in different locations, offbeat and a dramatic break from the orthodox university medical curriculum that was built on canonical texts. In his unconventional scholastic process, he became absorbed with astrology and alchemy that strongly influenced his idiosyncratic approach to healing.

His biggest heretical stance was that he thought that there was no greater obstacle to progress than traditional books. He believed texts should be discarded and that young physicians should return to nature and experience to become successful healers. Paracelsus preached that doctors must learn at the bedside of their patients, rather than spend time in libraries. This fostered a clear contempt for academic pomposity. He said, "I tell you, one hair on my neck knows more than all your authors, and my shoe-buckles contain more wisdom than Galen."

In the end, like his father, he took up the practice of medicine without the benefit of any academic credentials. He gained many years of experience as a military surgeon where he found that wounds healed better if they were not treated with traditional ointments. He concluded, "If you prevent infection, nature will heal the wound all by herself."

Afterwards as a civilian, Paracelsus lived in northern Europe and Basel. In his travels to other countries he absorbed the information by barber-surgeons, midwives, and folk healers. He adopted many of their unorthodox practices. This along with his strong religious belief led him to argue for a new approach to medicine.

Paracelsus from his alchemy background believed that God had created humans by using natural substances like salt and other materials. Therefore, God provided all the cures for illness in native materials such as herbs, vegetables and minerals. He advocated that physicians should use commonplace cures and reject Galen's theory of four humors as an explanation for illness and the treatments that followed from that hypothesis.

Paracelsus took the first definitive steps away from Galen's writings and in the process made a number of significant contributions to medicine including:

- He followed Hippocrates' observation-based medicine and that each disease was a separate entity that resulted from agents outside the body. This was the initial measure on the path to the germ theory discovery in the 19th century.

- His study of alchemy led him to introduce the idea that medicines could be extracted from various metals like arsenic, lead, sulphur and others. He knew that these substances could be poisonous, and the secret was in the dosage that makes it toxic or not. This paved the way for the serious application of chemistry to medicine. Because of this Paracelsus is often cited as the founder of toxicology (the study of the harmful effects of toxins or poisons on living things).

- Raised in a mining community, he recognized that the miners' lungs and skin absorbed noxious materials. This observation was a break through well before the industrial age in which the hazards of manufacturing became known.

- Visiting Constantinople Paracelsus learned how peasant women prevented smallpox by inhaling the pulverized scabs of smallpox lesions. This process was a full two centuries before inoculations were introduced in England.

- He was the first to manage effectively the congenital form of syphilis by using mercury.

- He believed that doctors should treat the rich and poor alike. Paracelsus advocated a graduated fee system to pay practitioners. The poor would be treated for free while the wealthy paid more for their medical care.

Paracelsus' novel theories and remedies were predicated on a rejection of ancient medicine. His work highlighted the divide between the old dogmas supporting the universe as a whole and the new ideas that appealed to many patients as well as those doctors willing to challenge the past. This contrarian approach resulted in fierce criticism by influential university professors and academics. This reception is not surprising considering Paracelsus' abrasive personality. He never learned the art of persuasion or embraced the Dale Carnegie philosophy of "how to win friends and influence people."

Paracelsus died in 1542 poor and scorned, even by his few friends. Yet he prophesied correctly that the massive Galenic edifice would collapse taking with it the Latin spouting doctors he so despised. This prophecy would take time to come true because few of his medical writings were published before his death. When they became available, Paracelsus' ideas spread in a blaze of controversy just like his life.

Ambrose Pare

Surgical procedures in the Renaissance were high risk. As in the Middle Ages they were left largely to the lowly barber-surgeons. For them, the new anatomical discoveries were of immediate practical value particularly with the introduction of gunpowder and canons that caused wounds to be wider and deeper. This led to widespread infection. Surgeons had a perilous new challenge that could not be solved by the medical theories of the ancients.

Into the fray came the greatest Renaissance barber-surgeon, Ambroise Pare (1510-1590). He trained with his father and others who were barber-surgeons. After completing his apprentice-based education in 1536, he immediately joined the army. Over the next thirty years he rotated in and out of private practice returning to military service when needed. He earned his first laurels on the battlefield that subsequently led him to question many established surgical procedures. Pare was well aware that surgery was dangerous. He only resorted to it when it was absolutely necessary. His willingness to experiment for better outcomes was coupled with a sincere compassion for his patients that put Pare in the forefront of change during his lifetime. His well quoted expression, "the art of medicine is to cure sometimes, to relieve often, and to comfort always" says a lot about this man.

A pivotal event occurred for Pare in 1537. He was serving as an army surgeon during the Siege of Turin. He ran out of the boiling oil concoction that was used at the time to sear and seal gunpowder wounds. This process allegedly "detoxified" the body of poisons that were believed to be carried by the gunpowder and the projectile. In need of an immediate alternative, Pare recalled an ancient treatment. He mixed a potion of egg yolks, rose oil, and turpentine and applied it to the soldiers' wounds. The next day Pare saw that the injuries were beginning to heal. Moreover, the horrific pain cause by the boiling oil treatment had been avoided.

In light of this experience, Pare resolved to change his attitude towards medicine and surgery. He decided to observe carefully, use his own judgment, try new ideas, and assess the results. This experimental approach went against the blind acceptance of age-old methods used by most physicians and surgeons of the time.

Throughout his career he implemented many advances. He wrote at length about his experiences in French rather than the usual Latin of medical textbooks. This allowed less educated barber-surgeons to learn new things about their profession.

Pare's many innovations included:
- He invented an early hemostat clamp (he called it "crows beak")—a version is used today.
- One of his greatest accomplishments was the reintroduction of the ligature. It had been abandoned since antiquity and replaced by cautery that the Arabs applied as a means of hemostasis.
- While he still used cautery for some circumstances, he rejected acid treatments to burn the wound and stop bleeding.
- In obstetrics, he reintroduced a method to rotate a fetus in utero to the headfirst position in order to ease delivery.

- Pare was sensitive to the experiences of his patients. If there was time or opportunity, those under his care were given wine, opium or henbane (a plant that contains scopolamine) to deaden their pain.
- From a female practitioner, he learned to use chopped raw onions to heal wounds. In the mid 1950s scientists analyzed this treatment and discovered that onions do in fact contain an antimicrobial agent.
- He learned to do a proper herniotomy. In those times, hernias frequently strangulated the bowel which if not decompressed often caused death. Pare's surgical method saved many lives.
- He developed limb prostheses as well as false eyes and noses.

Many of his contemporary medical experts recognized and accepted Pare's abilities and innovations. He helped raise the status of barber-surgeons. As a consequence, their profession gradually merged with those who had more formal training in surgery. Pare's talents were honored with an appointment to the French royal court. This secured his finances and gave him more time to experiment.

The legend of the bezoar stone is a fascinating tale that reveals another side of the innovative Pare. In 1567, Ambroise Paré described an experiment to test the properties of bezoar stones. The stones were commonly believed to be able to cure the effects of any poison, but he believed this was not true.

It happened that a cook in Paré's court was caught stealing fine silver cutlery. He was condemned to be hanged. The cook agreed to be poisoned, on the conditions that he would be given a bezoar immediately after the poison was ingested and that he could go free if he survived. The stone did not cure him, and he died in agony 7 hours after being poisoned. Paré had proved that bezoars could not cure all poisons.

It is hard to appreciate in today's democratic age, the tremendous implications of the rise of Paré , a lowly barber-surgeon, to such social and scientific heights in the then very highly stratified society of the Renaissance world. Only because of his high intellectual endowment, coupled with relentless labor and study along with the great strength of character allowed Paré to achieve such a result. The humility of the man is even more striking. At the end of his career, Paré would say, "I bandaged them, but God he healeth them."

Other Surgical Leaders

While Paré was a leader in his field, the great need for surgical expertise meant that others made progress as well. England participated in the renaissance of surgery through the works of Thomas Gale (1507-1586), William Clowes (1540-1604) and Gaspare Tagliacozzi (1546-1599).

Gale is noted for his active crusades against charlatans. In the military, he observed that some soldiers were being treated by cobblers and tinkers (the British name for hobos or tramps).

They were making things worse to the point that even those with minor wounds ended up dying. In civilian life, he saw similar incompetency where counterfeit doctors robbed patients of their money and also their limbs and perpetual health. Gale's campaign against this ineptness led to some success in reducing the number of untrained surgeons.

William Clowes became known as one of the most experienced surgeons in caring for men in the active military. His book about the treatment of wounds became a classic for military surgeons to us.

Italian Surgeon Gaspare Tagliacozzi (1546-1599) was an innovator. He revived the art of rhinoplasty that was first introduced by ancient Indian healers. This form of plastic surgery for the nose involved twisting a flap of skin from the forehead to come down over the nose in order to repair or change its appearance.

Surgery during the renaissance advanced the profession academically. But what is sometime forgotten is that many surgeons out of the limelight worked humbly in towns throughout Europe helping many patients. The kinds of operations that are routine today—like removing the appendix—were still impossible to perform by Renaissance doctors. Surgical procedures would continue to be perilous for centuries until better anesthesia, a fuller understanding of anatomy and physiology and a cure for postoperative infections became realities.

Midwives and Obstetricians

Women as midwives remained in charge of childbirth, except when a surgeon was called to extract a dead fetus from the uterus. Midwives, according to one Renaissance contemporary, "must be mild, gentle, courteous, sober, chaste, patient, wise, and discreet." While most served as an apprenticeship under an experienced midwife, their skills and knowledge varied. In England, the church licensed midwives. To be sanctioned they had to swear an oath to make their services available to both rich and poor and not to engage in magic.

Some English surgeons who specialized in difficult deliveries were predictably critical of their skills. The doctors bemoaned the midwives' tendency in a slow labor to tug upon a foot or hand, sometimes yanking it off altogether in the process. The physicians' concern set in motion the rise of man-midwives that as expected was criticized by women performing midwifery. Some surgeons began to add midwifery to their own practices and to attend to births, even when there was no medical emergency.

Not all women midwives were marginally qualified. There was an elite class of better educated midwives. This group is best represented by Frenchwomen, Louyse Bourgeois (1563-1636). She increased the level of professionalism among those who oversaw the birthing process. Her ethical precepts are still viewed as relevant today. She rose to this high level of respect in part by help from her husband. He was a barber-surgeon who shared with her his knowledge. Subsequently, she sought out what had been written about childbirth.

Bourgeois began attending the births of the working class where both her experience and reputation grew. She progressed to receive special certification that was needed to attend to noblewomen or royalty. After she delivered the wanted male heir for Henry IV and five more children for his wife Marie, she was given the title of midwife to the queen.

In 1609, she published her knowledge of childbirth. It was one of the first treatises on midwifery ever written. Unfortunately, her successful career ended twenty years later when the wife of the brother of the King died in childbirth in which she attended. Bourgeois herself died in 1636, but her legacy continued with several of her children who took up medicine and midwifery. Partly thanks to her in 1631 Paris midwives began to receive some formal training. In 1679 Amsterdam midwives were required to attend anatomy and obstetric lectures before being allowed to assist in childbirth.

Obstetrical skills advanced with one important invention. This was the development of forceps that was introduced by a French Huguenot, William Chamberlain (1540-1596). The innovation was that the two halves of the tongs could be separated at their crossover point. This permitted each curved hollowed metal blade to be inserted separately into the woman's pelvis in order to grasp the baby's head allowing it to be gripped and drawn out.

An incredible facet of this device is that Chamberlain kept the instrument a family secret. The covert forceps were passed onto his son and two more succeeding generations. When called to a delivery, the Chamberlains supposedly hid the forceps in a box to preserve their trade secret. It was easy to conceal in the labor room, since modesty demanded that a sheet cover the pregnant woman. Eventually, their deception was revealed, and they no longer had a corner on the market.

CHAPTER THIRTEEN
Medical Progress in the
Second Half of the Renaissance

T he second half of the Renaissance period encompasses the entire 17th century. It occupies a unique position in the history of science. It is the century of the mathematician-philosopher Descartes, the physicist-astronomers Newton and Galileo, and of chemist Robert Boyle.

The medical record is equally brilliant. The previous 100 years had seen the rebirth of anatomy, surgery, and the application of chemistry to medicine. All these branches of medical science continued to advance during the 17th century. In addition, two consequential fields emerged: physiology (the branch of biology that deals with the functions and activities of living matter) and microscopic anatomy. These areas provided the icing on the cake of accomplishments that occurred medically in the Renaissance era.

William Harvey Discovers Blood's Circulation

The greatest physiological advance of the 17th century, and perhaps of all times, was the discovery of the circulation of the blood. The concept of circulation—blood pumped by the heart and traveling around the body through vessels—seems obvious today, but it was a mystery for thousands of years. No other aspect of the body has been subject to so much superstition or conjecture over the ages until the truth was finally revealed in 1628 by William Harvey (1578-1657).

He rebuked Galen's false theories concerning the movement of blood that said:

- There were distinct functions for venous and arterial blood.
- The heart did not pump out.
- There were mini holes in the septum of the heart.
- Nutrition was absorbed in the blood in the liver.
- And air from the lungs went directly to the left side of the heart.

In hindsight, the lack of a sound observational basis for many of these particulars appears problematic, but the Galenic cardiovascular system was embraced as truth for so long because it explained everything simply and coherently.

Enter English physician William Harvey. He performed experiments that overturned centuries of speculation about the veins, arteries and the workings of the heart. He was the

eldest of nine children born in the southeastern part of
England. His father was a successful businessman, and his
brothers followed into the family business. William elected to
become a doctor. He studied at Cambridge and then medicine at the University of Padua where he graduated in 1602.
Shortly thereafter, Harvey married the daughter of a prominent English physician who helped him secure a position at
London's St. Bartholomew's Hospital and as a fellow of the
Royal College of Physicians.

Harvey was described by one 17th century contemporary observer as a swarthy (of dark complexion) and testy "hot-
headed" young man. He habitually wore a dagger that was
ready for use at the slightest provocation. His ill-tempered
personality did not stop his advancement, and by 1618 he was
one of the royal physicians.

Harvey's experience at Padua set him apart from other doctors and put him in the forefront of learned medicine. He was
a man who admired Aristotle and valued the views of Galen
but was not afraid of challenging the old master. In the end,
Harvey relied on his own observations and reasoning to
develop conclusions.

In 1628, he published *De Motu Cordis* (On the Motion of the
Heart and Blood). It was a slim 72-page book that capsulized
more than 20 years of dissections, observations and experiments on over 60 different species of dead and living animals.
The circulation of the blood was not merely put forth as a
theory. It was proven by morphological, mathematical, and

experimental arguments. It was the first well rounded scientifically established description of the circulation system.

Harvey showed, what every grade school student knows today, that the heart is a muscle and that there is a one-way system through the arteries and veins. He further demonstrated that the blood picks up a fresh supply of oxygen in the lungs and returns it to the heart where it is then pumped out into the body. In the process, he undertook the first quantitative study of blood flow. He accurately measured how much blood passed through the heart each day. This along with many other detailed scientifically designed experiments was the foundation for his breakthrough findings.

In spite of his thoroughness, there was one major gap in the circulation route that Harvey was unable to close. He could trace the blood through smaller and smaller arteries and veins to the limit of the unaided human vision. But without a microscope, he could only guess that invisible pores were the mechanism that the smallest arteries connected with the veins. It was not until 1661, after Harvey died, that Italian biologist Marcello Malpighi finally detected under the microscope the capillaries that couple the veins and arteries. Five years later, minute red corpuscles were observed streaming through the capillaries. The circle that Harvey began was now complete.

Harvey's discovery encountered violent opposition not only because it was anti-Galen, but also because it went against the theory and popular practice of bloodletting. As a sign of those turbulent years, Harvey's London medical practice suffered. With time good sense prevailed. Physicians and surgeons accepted his findings well within his lifetime. Medicine would never be the same again, and patients would be eternally grateful.

Even with his magnificent discovery, Harvey realized how much needed to be learned. He wrote, "All that we know is still infinitely less than all that still remains unknown." That continues to be true today.

Other Physiological Advances

Circulation was not the only aspect of physiology to develop in the 17th century. Harvey lived long enough to see others build onto the knowledge. He inspired important observations concerning respiration and digestion.

Robert Boyle (1627-1691) revealed that animal life is not dependent on air in general, but rather on one particular component of air. A century later a fellow Englishman Joseph Priestley discovered that component was oxygen. Robert Hooke (1635-1703) showed experimentally that the mechanical movement of the thorax was not the essential element in respiration. He demonstrated life could be maintained in animals even after the thoracic wall was removed by using bellows to blow air through the trachea into the lungs. Jean-Baptiste van Helmont (1577-1644) described digestion as a series of fermentations involving hydrochloric acid that he identified in the stomach.

The discoveries in digestive and respiratory functions did not rise to the same level of advancement as the physiology of circulation. This was because the underlying problems of these two areas were chemical in nature, and chemistry had not yet been fully developed.

There is interesting side light about a well-known 17th century English astronomer, mathematician and architect. While Christopher Wren (1632-1723 is best known for his buildings, his other endeavors included microscopy, optics, surveying, meteorology and medicine. In 1665, Wren was participating in a medical experiment with canine transfusions. While giving a dog a shot of opium using a bladder attached to a sharpened quill, he realized that injections could also apply this technique intravenously. But it would be much later that an IV was a common medical procedure.

Clinical Breakthroughs Overshadowed By Pure Science Advances

By the last half of the 17th century, it was apparent that the new basic sciences of physics and chemistry had little application to clinical medicine. As a writer in 1680 noted, "All in his profession agree that Dr. Harvey was a revolutionary scientist, but I never have heard any that admired his treatment of the sick. I knew several doctors in London that would not have given a three pence for one of his prescriptions."

Outside of academic pursuits, everyday medical care remained largely faithful to the old heritage. Patients were conservative and preferred the therapeutic devil they knew. Into this confused environment, advancements for the populace did make some breakthroughs. This headway is often overshadowed by the purely scientific progress of the period. The theories of Galen were out, but Hippocrates maintained his Olympic status as the champion of the bedside experience.

The best physician representative of this approach was Thomas Sydenham (1624-1689) who sometimes was called the "English Hippocrates." He was a practical man free of the sterile theorizing of many of his contemporaries. Sydenham was in the army before he studied medicine relatively late in life. He received his degree from Oxford at age 39. He practiced in London while investigating smallpox and other fever like maladies. Revering Hippocrates, Sydenham's clinical method was simple. Medicine was a craft that would progress only through observation of patients and monitoring of therapies. He said, "You must go to the bedside. It is there alone that you can learn disease."

Sydenham was one of the first to recognize that different afflictions were distinct and possessed unique natural histories. He emphasized how each ailment was independent of the individual patient. He was known for his treatise on gout, of which he was a sufferer. Sydenham was famous for his studies of malarial fever, dysentery, measles, scarlet fever and chorea minor that bears his name. Sydenham Chorea or St. Vitus dance is a rare autoimmune neurological disorder of children and adolescents characterized by rapid, involuntary, purposeless movements especially of the face feet and hands. There is no known cure.

Sydenham's individualism of diseases made him sympathetic to the advanced idea of specific remedies for particular disorders. His adoption of one of the few effective remedies that appeared at this time is illustrative. The medication was imported from Peru beginning in 1630 and called "the Jesuit powder." It cured one of the most common diseases of the period, malaria. It came from the bark of the cinchona tree that was later shown to contain quinine. It was well received because it did not produce any of the "evacuations" claimed necessary by the Galenists and humoral pathologists. Sydenham's philosophy undermined to a considerable extent, the traditional pharmacological and pathological theories. This is one of his enduring legacies.

While there were other creative clinicians in the 17th century, the average product of universities brought no revolution in medical services or treatments. These ordinary doctors tended to exhibit limited learning and clinical skills while utilizing the harmful routine of abundant purging and bloodletting in therapeutics. Their medical strategy reflected that the universities in general were backward, almost medieval, and not attuned to the scientific progress of the time. Practically all the great discoveries in the last years of the Renaissance were sponsored by various academies and learned societies, and did not originate within the university setting.

The Rise of the Microscope and Microanatomy

Germs, tiny parasites, and the smallest blood vessels that all
are invisible to the naked eye were unknown to Renaissance
physicians before the invention of the light microscope. This
simple devise revealed the world of the very small. It began
the field of microanatomy that explored afresh the human
body. The development of the microscope was one of the
most significant inventions of the period. Over time it had a
major effect on medicine by giving scientists the capability of
seeing things that previously they could only imagine.

While the rise of the microscope is connected to Anton van
Leeuwenhoek (1632-1723), simple magnification lenses were
seen as early as the Roman era. By the 13th century, Italians
had begun to create lenses that could be worn as eyeglasses.
Some modifications through the years continued, but a
magnification factor was never achieved greater than 10x
until the unlikely Dutch cloth merchant, Leeuwenhoek,
moved science forward in a major way.

He used simple magnifying lens in his commercial work as a
cloth merchant to ascertain the thread count of woven
materials. He became fascinated with the process and created
new ways to grind and polish very small lens by giving them
additional curvature that increased the level of magnification.
Leeuwenhoek was very sensitive to capturing the light when
using the lens. This greatly enhanced what he could see. He
was able to achieve a magnification that enlarged things 275
times bigger than normal.

The basic light microscope is still of value in medicine. There are always objects that do not require super powerful magnification. Today scientists benefit from seeing what would have been unthinkable in Leeuwenhoek's day by using the electron microscope. This employs beams of electrons focused by magnet lenses instead of rays of light. Its magnification is significantly greater than that of any optical microscope.

With his microscope, Leeuwenhoek went from looking at cloth to gathering various animals, plants, and mineral crystals and observing them under his invention. He was remarkably dexterous in preparing samples. As a tradesman without formal scientific education he had a mind free from any academic prejudice. In water from ponds he saw what he termed "animalcules"—in modern terminology, a single celled microorganism. He was the first to see roundworms, blood cells, sperm cells, and striated skeletal muscle cells. Many of his numerous discoveries were of medical importance.

Leeuwenhoek worked meticulously documenting what he observed. He had a professional artist illustrate his findings. Eventually, his work made its way to the Royal Society in London where he began a long series of correspondence detailing his microscopic observations. All his letters were written in Dutch, and he never published the material in book form. Scientists soon realized what this tradesman was doing was indeed quite remarkable. He became famous all over Europe during his lifetime and with his death gained a page in the history of medicine. Interestingly, none of the lenses he created never surfaced after he died. Because Leeuwenhoek used gold and silver to make his instruments, his family likely sold them.

A contemporary of Leeuwenhoek was Italian scientist, physician and anatomist Marcello Malpighi (1628-1694). Malpighi was the principal architect of microanatomy. He laid the foundation for a new branch of science, histology. Derived from the Greek word *histos*—meaning web or tissue—histology is the study of tissues, which are a collection of similar cells, such as muscle, bone, nerve, or cartilage.

He studied different kinds of plants and animals near his estate in Bologna. His identification of capillaries in frog lungs in 1661 was the missing link between arteries and veins in the circulatory system described by William Harvey. Malpighi devised new methods to illuminate tiny specimens by staining them so they could be better appreciated under the microscope.

Like so many innovators, his discoveries challenged the approaches and beliefs current at the time. He provoked controversy that made him unpopular among his colleagues. Malpighi died of a stroke in 1694 while serving as the physician to the Pope in Rome. His name is commemorated to the present in many areas of biology and microanatomy.

Robert Hooke
A Forgotten Genius and True Renaissance Man

Robert Hooke (1635-1703), the man who coined the term "cell" to describe the basic unit of life, is a forgotten genius. He is the appropriate person to complete the discussion of medicine in this era. He could well be the template of what is a true Renaissance man.

Some consider him the single greatest experimental scientist of the 17th century. Like the 14th century Leonardo da Vinci, he delved with expertise into so many fields including physics, astronomy, chemistry, biology, geology, architecture, and naval technology. Among his accomplishments were:

- He created a balance spring that increased the accuracy of timepieces.
- He devised the theory of elasticity that is still used today. (The property of solid materials to deform under the application of an external force and to regain their original shape after the force is removed is referred to as elasticity.)
- He helped develop the vacuum pump with chemist Robert Boyle.
- After the devastating Great Fire of London in 1666, he worked as surveyor and architect to rebuild the city. Hooke collaborated with Christopher Wren on St. Paul's Cathedral, whose dome uses a method of construction conceived by Hooke.

- In his theoretical study on gravity, he sparred with Newton.
- In astronomy he was an early observer of the rings of Saturn.
- He theorized about how living things could be turned into fossils by mineral-rich water washing over them, leaving behind mineral deposits over a long period of time. Hooke's idea was 250 years before Darwin concluded that fossils are not accidents of nature, but the remains of once living organisms.
- One lesser known, but ahead of his time, theory was his scientific depiction of human memory. His model addressed the components of encoding, memory capacity, repetition, retrieval, and forgetting—with surprising modern accuracy. And this was done in 1682.

Though Leeuwenhoek was referred to as the father of microscopy, Hooke could also be given that mantle. He created a compound microscope and illumination system that was one of the most advanced at the time. His extensive studies and detailed illustrations of his microscopic observations were published is a 1665 book, *Micrographia*. It became a best-selling monographs on science. To make it accessible to as many as possible, Hooke wrote it in English not Latin.

In his publication Hooke portrayed tiny objects from insect bodies, eyes, and legs to bits of blossom, seeds, and other plant specimens. Hooke's legacy was established when he became the first scientist to use the word "cell" to describe tiny boxlike compartments in plant specimens of cork. He likened them to the rows of similar shaped spartan chambers known as cells occupied by monks. The term soon passed into general usage as the word that defined the smallest living components of plants and animals.

Modern historians acknowledge his preeminence as a Renaissance scholar extraordinaire, but shortly after his death he was all but forgotten. Isaac Newton had become president of the royal society of which Hooke had been a prominent member. Newton had vehemently opposed Hooke's views on gravitation that he had presented to the society. Though the true story of what actually happened is not known, some speculate that Newton obscured the work of his arch rival scientist Hooke because of their disagreement.

One sad fact for medical historians is that there is no portrait of this great Renaissance Man. The only rendering of Hooke was a stained glass window in an old London church that was destroyed in a 1993 in an IRA bombing. His microscope, however, is displayed in Washington at the National Museum of Health and Medicine.

Some Conclusions About Renaissance Medicine

It is difficult to draw conclusions about Renaissance medicine because what occurred then was almost schizophrenic in nature. During the first one hundred years after the Middle Ages, humanism's dream and preoccupation was to restore medicine to its Greek purity. This approach was a mixed blessing and skepticism towards physicians remained deep-seated. Shakespeare said it best, "trust not the physician, his antidotes are poison."

During the 17th century a robust transformation in medicine is some circles began to build on a new scientific basis. This counter view gained ground as many doctors refused to pay homage to antiquity. After centuries, the deadweight of the past started to be cast away. Adherence to Galen's teachings was a subversive doctrine for those change minded medical professionals. But in reality, what these reformers advocated was no different than what was happening in multiple other areas in society during that era. As one author said, "If Luther could break with Rome, how could it be irreverent to also demand the reformation of medicine?"

PART III
THE EMERGENCE OF
MODERN MEDICINE

In the three hundred years from 1700 to 2000, medicine emerged from a primitive science of longstanding traditions based on ancient theories and unproven remedies to modern diagnostic techniques and treatments. Quite literally more medical discoveries were made in these three centuries than in the previous 3000 years.

Think of it this way. If in 1700 a physician made a house call, the doctor's small bag could carry the entire diagnostic and therapeutic measures known at that time. Today, the physician needs a whole building to house all the modern medical procedures to find out and then cure an illness.

Even with the fundamental advances during the Renaissance, medicine was still a young science at the beginning of 1700s. It was not until the mid 19th century that the existence of germs became known. Prior to then, solutions that healers might have tried could not address the root cause of most illnesses. While surgery existed, it was a gory painful affair with high mortality and morbidity until anesthesia and aseptic techniques were developed.

The treatment of the mentally ill was often both cruel and ineffective until the second half of the 20th century when therapeutic drugs became available. In 1700, most public health measures were largely based on speculation and often self-defeating until the late 19th century. Medical education of physicians took a long time to fully integrate new scientific discoveries into clinical practices. Sentinel scientific innovations and those responsible for these discoveries demonstrate the collective impact of new ideas on society and how the average citizen ultimately benefited from these changes.

The topics in this part are divided into three segments that closely corresponds to each century. The first section is the 18th century. It has been called the adolescence of modern medicine. During these 100 years the early groundwork of scientific medicine was first established. This period is characterized by the ideas of the philosophical movement known as The Enlightenment. It inspired the search for rational systems of medicine, practical means of preventing disease, improving the human condition, and disseminating the new learning to the greatest number of people possible.

The second segment starts in 1800 and concludes at the onset of World War I. During this time, the conceptual foundation and practice of medicine were slowly transformed into a modern design that still characterizes the profession today. There were parallel developments in biomedical sciences, physiology, pathology, immunology and other areas. New discoveries slowly reshaped how doctors practiced medicine. Previously, the profession often ignored the scientific advances and based their clinical work on custom, tradition, and individual experience. By 1909 medicine was so intertwined with science that it was set to explode even further on multiple fronts.

The 20th century had two phases of development. Growth in fundamental scientific discoveries and clinical applications occurred from the initial decade until the end of World War II. After 1945 there were expediential modifications in the prevention, diagnosis, and therapeutic cures for multiple diseases and conditions particularly in the United States.

One author's comments about American physicians' and scientists' contribution to the transformation of medicine during these three centuries are of interest. Until the end of the 19th century there was a paucity of early innovators from the United States. American doctors were more often consumers than producers of scientific ideas. European scientists dominated the new thoughts until after World War I.

This observation is validated by the fact that United States physicians and scientists were shut out of the Nobel Prize in Physiology or Medicine for its first 33 years of existence— only Europeans were honored. It was not until 1934 that a team of American researchers won the award for the first time. Since then Americans have received the honor numerous times. After 1934 the United States was no longer considered a hinterland in medicine.

CHAPTER FOURTEEN
18TH Century—The Age of Enlightenment

Historical scholarship usually assigns the beginning of the Age of Enlightenment to the year 1688 in England when the Glorious Revolution occurred. This was also called the War of English Succession in which James II was replaced by William and Mary. There is debate on whether the Enlightenment ended with the Declaration of Independence by the United States in 1776 or the French Revolution in 1789. This period is best characterized as the time when the educated and professionals in Europe were willing to truly question old standards on everything within their society.

The central feature of this movement was the promotion of the power of reason as a means to free humankind from superstition and religious authoritarianism. Intellect and rationality would bring a better future to all historic issues confronting mankind. A major component of this philosophy was that science and technology would enhance man's control over nature. Social progress, prosperity and the conquest of disease would then follow.

Enlightenment thinkers looked to science as their model for improvement because it was objective, critical, and progressive. This spirit encouraged medicine to search out the wider laws of health and sickness. As a result, a much more optimistic outlook concerning the role and benefits of medicine developed. In the end, while the Enlightenment era saw a limited number of positive scientific advances to help the ill, it more importantly began to reshape the perception of medicine's place in society setting the stage for progress in the future.

18th Century Medical Systems and Theories of Life

Beginning in 1700 scientists attempted to systematize simple fundamental principles pertaining to the various functions of living beings in health and sickness. Enlightenment physicians sought a rational theory of medicine that could provide comprehensive explanations of disease causation that would become a firm foundation for medical practice.

Their intellectual goal was to achieve a straightforward and logical synthesis of medical knowledge designed to replace the increasingly obsolete humoral concepts inherited from antiquity. Recall that it was Galen, the Greek physician who lived during the Roman Empire, who pioneered this theory. For 1500 years succeeding generations of doctors believed and embraced his hypothesis that the basic composition of sick persons was an unbalance of 4 humors (blood, phlegm, yellow and black bile). In the 19th century this ancient conjecture was finally put to rest.

There were a number of different medical systems proposed in the 18th century about the origin of diseases and how normal functions in the human body worked. One theory was designed by Dutchman Hermann Boerhaave (1668-1738). He was considered the most successful clinician and teacher of the 18th century. Along with his great emphasis on bedside instruction, his eclectic approach to challenges, and his charismatic personality attracted wide attention. As a true humanist in the Renaissance mold, his interests wandered far beyond medicine to encompass all the arts including music and literature. His contemporaries thought of him as the 'Newton of Medicine.'

Boerhaave promoted a theoretical medical system that incorporated the most important physical and chemical advances made during the preceding century. He saw health in terms of hydrostatic equilibrium, a balance of internal fluid pressures. He believed that disease occurred when the normal interaction of circulating fluids was disturbed. Unfortunately, he made the time worn therapeutics inherited from Galen appear to be logical derivations from his theoretical scheme. Bloodletting was designed to reduce the flow and volume of blood, thereby reducing its deleterious pressure on certain obstructed blood vessels. Purging was beneficial to shrink bodily fluids when there was flooding of critical organs.

Boerhaave's hydraulic model of the body incurred criticism showing that 18th century medicine was far from monolithic. Many of his students became noted scientists throughout Europe and developed competing theories.

German George Ernest Stahl (1659-1734) maintained that human disease and actions could not be solely explained in terms of mechanical chain-reactions as was seen by Boerhaave. Instead Stahl advocated an 'animism,' a super-added soul, as the prime mover of living beings. This 'anima' was the agent of consciousness and physiological regulation. Disease, Stahl said, was the soul's attempt to expel morbid matter and reestablish bodily order. Animism was actually nothing new. It was simply a revival of the ancient idea of benevolent nature that must be allowed by the doctor to take its course.

Another German Frederic Hoffman (1660-1742) promoted his own system. He emphasized the role of nerves in physiology and pathogenesis. For him he theorized that small particles flowed through the nerves creating a degree of tension or tonus in all bodily fibers that ensured healthy functions. Disease for Hoffmann was the result of abnormal bodily motions precipitated by changes in the nervous system. Based on his hypothesis relatively simple therapy consisted of administering relaxing sedatives or irritating stimulants were the best methods to treat illnesses.

One of the most notable of Boerhaave's disciples was Swiss Albrecht von Haller (1708-1777). He was a man of unlimited energy and imagination. He expanded the theories with some elementary experimentation saying that irritability or contractility was a property inherent in muscle, whereas sensibility or feeling was the exclusive attribute of nerves. Henceforth, as a result of his supposition greater emphasis was placed on the role played by the neuromuscular system in the phenomena of health and disease.

In the English-speaking world, the most influential attempt to set disease and bodily functions in a coherent framework lay in the teachings of Scotsman William Cullen (1710-1790). Like Hoffmann, but with a stronger emphasis, he believed that the nervous system was the key to physiological balance. For Cullen, life consisted of a state of nervous excitement or irritability produced by environmental stimuli. Rejecting humoralism, he held that all pathology originated in a disordered action or spasm of the nervous system—earning Cullen the nickname 'Old Spasm.'

Early in the 18th century the medical systems conceived and speculated by Boerhaave, Stahl, Hoffman, von Haller, Cullen and others brought some stability to the thinking of the day on the theory of life and disease. This feeling was short lived and its downfall was inevitable. These modes of cognition filled scholars with enthusiasm at the time, but no single framework provided medical certainty or advanced the practice of medicine.

Most physicians of the era gradually recognized that much more basic knowledge was needed before a genuine synthesis could be found. All systematizing efforts were therefore premature. In ever increasing numbers 18th century physicians abandoned the armchair scientists and returned to the sickbed to obtain knowledge. For many, the collection of clinical facts became the proper road toward medical certainty and the understanding of the underlying disease process occurring in their patients.

Disease Classifications: Nosology and Early Pathology

With the medical theories and ideas in dispute, doctors hesitated to invoke unknown speculative causes for disease. Instead, they built a new system of cataloguing diagnoses based on the careful observation of symptoms. It was a self-conscious effort to avoid theorizing and bring order to clinical medicine. It was called nosology. This is defined as a "disease classification" or as one author calls it, patterns of suffering. It was derived from the Greek words for disease and theory about.

The roots of nosology originated in the 17th century physician Thomas Sydenham (1624-1689) who was known as the English Hippocrates. His publication of clinical observations about disease, especially fevers and their treatments, began the movement. Enlightenment physicians throughout Europe embraced the concept and nosology became the new spirit of the times. It evolved into the early framework for pathology as a system of medical knowledge. Many different classifications were developed by various doctors.

The construction of a new nosology was a complicated process. Each disease was clinically enumerated by its essential and constant symptoms and signs. Particular sequences were lifted out of the confusing array of clinical events and defined as individual diseases. Much of this data was acquired in the subjective reports of patients that were obtained in the form of clinical histories. The physician searched for objective evidence at the sickbed. This was a rather undeveloped art in the 18th century because the physical examination was restricted to inspection and palpation.

The practitioner had to discriminate between a number of seemingly similar manifestations and recognize distinctive, so called pathognomonic events, which at times were considered specific for a certain disease. The next phase of nosology's development demanded that practitioners compare the collected information on the various sicknesses among themselves looking at similarities as well as different features. Each physician then collated the diseases and separated them into various families or classes based on anatomical, symptomatic and pathological criteria. The result was an organized framework for all diseases known at the time.

This process for the first time introduced orderliness into the practice of medicine. It represents an important effort by contemporary physicians to discern separate disease states and systematically survey the perceived panorama of clinical pathology. Physicians saw the various classifications produced as true reflections of an established natural order. This development was necessary to achieve successful diagnoses. It was

the first step physicians needed to begin their search for specific more effective cures.

Nosological classifications are still used today in pathology
and clinical medicine. It simplifies the mass of information
accumulated in clinical experience by giving it order and
structure and to serve judgment on diagnosis and prognosis.

In contrast with 18th century, most modern nosological systems now in place refer to established objective anatomical or
chemical changes in diseases. Interestingly only in psychiatry,
where specific physical or anatomical lesions are usually absent, do doctors continue as in the 18th century with a similar
ordering of knowledge, based only on subjective observational symptoms and behavior.

Humans have always tried to understand illness, injury and
death. While the nosologists gathered, sifted, and classified
illnesses for practical purposes, 18th century anatomists used
the postmortem exam to study the bodily changes brought on
by disease. Although they were not called so at the time, these
scientists were the first pathologists. They recognized the
connection between the sick person and the disease signs
afforded by the corpse.

This new approach to the autopsy was pioneered by Italian
Giovanni Battista Morgagni (1682-1771). Many historians cite
him as the founder of modern pathology. His three-volume
masterpiece of pathological anatomy, *On the Seats and Causes of
Disease* (*De sedibus et causis morborum*) was based on over 700

dissections. It was published in 1761. The book was the first to establish a retrospective link between lesions found in the body and clinical symptoms of the patient that died.

He made pathological anatomy more accessible to clinicians by including an index for diseases and another for specific lesions. He hoped the reader's learning of one might lead to knowledge of the other. Morgagni's emphasis on the explanation of disease shifted the thinking from the antiquated concentration on general conditions and humors to the more appropriate study of localized changes in organs and their connection with clinical symptoms.

Morgagni's discoveries were numerous. He was the first to notice fibrous clots in the heart as a result of cardiac disease. He observed arteriosclerosis in coronary arteries, and that stroke or apoplexy—as it was called then—was caused by an alteration in cerebral blood vessels. He grasped that when only one side of the body is stricken with paralysis, the lesion lies on the opposite side of the brain. He also accounted for aspects of gastric ulcers, abnormal vermiform appendix, emphysema and other conditions.

After his book's publication colleagues offered him bizarre reports of what they speculated were causes of various symptoms. In one example, discussing fluxes or diarrheas Morgagni viewed with skepticism all accounts and reports that the ingestion of the excretions of frogs, toads, or lizards was the cause for loose stools. Good call Doctor Morgagni!

Morgagni was a trailblazer of morbid anatomy and served as the initial guide to a new epoch in medical science. Even though he remained essentially a humoralist, his work marked a trend away from that theory towards the study of localized lesions and diseased organs. His work helped establish an anatomical orientation in pathology. Morgagni recognized that unseen anatomical changes in the body were often reflected in the physician's clinical evaluation of the patient. Unfortunately at the time, the doctor's bedside conclusions could only be confirmed in the autopsy room, not while the patient was still alive.

Others recognized the great significance of Morgagni's work. In France, the 1799 publication of Francois Xavier Bichat (1771-1802) shifted the focus from Morgagni's concentration on complex organs to that of individual tissues. He described 21 types—like connective, muscle, and nerve tissue—distinguishing each by appearance and vital qualities. They were the building blocks and provided a new map of the body. Henceforth disease was to be lesions of specific tissues rather than simply organs.

Bichat saw pathology with fresh eyes. His research formed a bridge between the morbid gross anatomy of Morgagni and the 19th century cellular pathology championed by the great Rudolf Virchow.

18th Century Therapeutics

The advances in pathology pinpointed a paradox. Prominent British physician and anatomist Matthew Baillie (1761-1823) said, "I know better perhaps than another man, from my knowledge of anatomy, how to discover disease, but when I have done so, I don't know better how to cure it." Therapeutics made herculean efforts, but the net contribution of physicians to the relief and cure of the sick remained marginal.

Most treatments were, more often than not, positively harmful. The fondness for heroic bloodletting and strong purges of calomel (mercurous chloride) was based on ancient theories. Nearly every physician's bag in Europe and in the large cities of America in the 18th and most of the 19th century had little 'blue pills' of calomel. Doctor Benjamin Rush of Philadelphia, patriot and founding father of American medicine, was the most sanguine advocate of copious bloodletting.

Yet, in spite of the detrimental effects of these popular heroic remedies, a prescription medication was an expected outcome by a patient after a medical consultation. Whether used simply or in compound mixtures, distilled, dried, or ground herbs constituted the bulk of the *materia medica*. Herbal remedies were largely designed to act as emetics and laxatives. By the end of the 18th century some new compounds and drugs appeared including castor oil, tartrate of iron and oxide of zinc. Opium was freely available over the counter and widely used in the liquid form called laudanum. It was taken as an analgesic, fever reducer, sedative, and for diarrhea.

The only truly effective drug in the doctor's bag for a specific illness was introduced in Europe in the 17th century. It was cinchona also known as Peruvian or Jesuits Bark. It was used to alleviate malaria symptoms and all kinds of other fevers. Today, we know that it contained quinine.

Letters and diaries show that when people fell sick they frequently framed their own diagnosis. For the increasing literate underclass, home medical manuals like "The Poor Man's Medicine Chest" published in 1791 brought simplified versions of elite medicine to the people. Common folks learned how to treat their ills with the aid of simple kitchen ingredients like onions and honey. Licorice was good for pacifying a consumptive cough. A deep cut's bleeding could be stopped by binding toasted cheese on the injury. Many turned to store bought items. In 18th century England ready-made medicine chests became popular. It was common for middle class families to stock up with patent medicines. While some were respectable, others were a product of quackery.

There were some significant innovations that were precursors to future medication. In 1763 "willow bark" was introduced. Salicin its active ingredient had a similar effect to aspirin that was introduced in 1897. In 1785 foxglove with its digitalis properties had a powerful stimulant action on the heart. It also reduced the edema or dropsy commonly associated with heart disease. While not a pharmaceutical in the true sense, in 1754 Scottish naval surgeon James Lind discovered the non-drug cure for scurvy. It was a nutrient in lime juice that was later named Vitamin C. It stopped the ravages of the disease.

It would be safe to conclude that overall the established 18th century pharmacy left much to be desired. This brought about a rise in quackery and irregular medicine.

The Golden Age of Quackery

The 18th century was arguably the golden age of quackery. The origin of this term is obscure. One author speculates it comes from the Dutch word quacksilver meaning a quicksilver doctor. At the time quicksilver was the name for mercury that was widely used to treat syphilis.

To speak of quackery is not automatically to impugn the motives of unqualified practitioners or irregulars. Many had some healing powers. They were not so much swindlers as entrepreneurs or fanatics about their techniques or nostrums. Take Scotsman James Graham (1745-1794), he touted that long life and sexual rejuvenation could be achieved by mud-bathing and use of his special electrified "Celestial Bed" housed at his 'Temple of Health' in London.

The majority of these charlatans were experts in the art of publicity. Vendors of "Rose's Balsamic Elixir" claimed it would cure beyond all other medicines any Englishman who was 'Frenchified'—in other words, one who had venereal disease. The product would according to its promoters remove all pains of the disease in three or four doses. In short, that elixir like many others was said to work like magic.

Most of the deceivers were small timers, but some made big killings. Joanna Stephens hawked a remedy that promised to dissolve painful bladder stones without the need for surgery. She was so convincing in her promotions that even members of Parliament raised five thousand pounds to buy the recipe from her. Patent medicines that promised to cure cancer, to restore lost youth, to increase potency, or to alleviate venereal disease were often successful because patients were frequently embarrassed to consult their family physician about these concerns.

One irregular medical approach that began in the 18th century that survived into the 20th century and still used today by some is homeopathic medicine. It was the brainchild of Samuel Hahnemann (1755-1843). His system evolved as an alternative to the heroic and often fatal therapeutic methods of the day of extensive bleeding, purging, and large doses of toxic drugs that induced vomiting.

Hahnemann advocated the use of infinitesimal small doses of such drugs as would produce the symptoms of the disease when given in larger doses. The system is summed up in the phrase, *Similia similibus curantur*—"like is cured by like." The theory has not been confirmed by scientific experience, but it was no more erroneous than any other 18th century therapies. Many more modern studies claim it is in fact the placebo effect at work—that is, if someone believes that they will get better, they have an increased chance of improvement.

With the rise of literacy, the public was eager to exercise its judgment and consumer power in the latter half of the 18th century. Demand welled up for all sorts of healing. Both government and medical authorities tried to clamp down on these irregular practices without much success. The more that was tried to limit or stop them, the greater their popularity became. This is best exemplified by Mesmerism.

Physician Franz Anton Mesmer (1734-1815) popularized an unconventional healing method. His checkered medical career began in Vienna. After graduating, he developed unorthodox healing methods almost by accident based upon his unconventional belief that all humans contained animal magnetism. The term animal as Mesmer used it, refers to the Latin word *animus* meaning breath or life force.

His inspiration came from a young woman patient in 1774 who was suffering from crippling seizures, vomiting, temporary blindness, and occasional paralysis. He had her swallow a preparation containing iron. Then Mesmer placed magnets over critical portions of her body. This resulted in convulsive responses and remission of her symptoms. Mesmer claimed that he restored the flow of her animal magnetism and in doing so brought about a cure.

Mesmer continued to treat other patients after simplifying the process. By just laying his hands on the ill person, he contended that it would transfer a magnetic force to the patient and restore health. His successes were likely due not to his personal non-existent magnetic force, but to the power of suggestion.

Mesmer's quackish theory and practice was not accepted in Vienna, and he was forced to move to Paris. Here, he procured an office in a wealthy neighborhood. Because he soon had a long waiting list of patients, he devised ways to treat groups of people collectively all at one time. He became a celebrity for his fashionable séances at which he cured nervously afflicted women through his animal magnetism. Patients would gather around a big tub filled with water and iron rods. A primitive battery was attached so that when one touched a rod they received a light electric jolt that was considered part of the cure.

The charade did not last. When rumors that his sessions were actually sexual orgies, King Louis XVI investigated. He put the famous American Benjamin Franklin in charge of the inquiry because he was well respected, understood science, and was in Paris as an ambassador from America. In the end, Mesmer was found guilty of charlatanism and was obliged to leave France. But the practice Mesmer founded continued in various forms by different practitioners for the next 50 years.

Medical Practitioners

If self-medication failed or the illness was serious and the patient could afford the services of a medical practitioner, there were three types from which to choose. The highest in

Professional status were university-trained physicians. Although they had no real cures, they provided a diagnosis. These top physicians were part of the elite and tended to concentrate in the large cities in Europe and America. The second level was the surgeon. More on this group follows.

When homecare failed many went to an apothecary, the third level of practitioners. This class filled a need by providing primary medical care for those who came to their shops. After some initial conflict with the physician elite, the apothecaries were eventually accepted within the medical community as general practitioners. This was an uneasy alliance that raised many questions of medical ethics. The public greatly benefited from the influence of these early druggists. The apothecary, today's pharmacist, persists in having an important role in today's healthcare system.

Meanwhile, surgery in the 18th century was undergoing changes in techniques, scope and status. For centuries surgery was considered a manual skill rather than a science. It carried little prestige. Barber-surgeons who learned their skills through practical experience, not a formal education, characterized it. These early medical interventionists day to day business revolved around minor procedures like lancing boils, dressing abrasion, pulling teeth, trussing hernias, treating skin lesion, and so forth. They did not attempt high-risk operations on the abdomen or complicated amputations. Remember this was the era before anesthesia and predates the profession's acceptance of aseptic techniques by a half-century.

Beginning in the mid 1700s European practitioners succeeded in emancipating themselves from their lesser half, the barber-surgeons. This occurred in large part because of the acceptance of surgical teaching as part of the overall curriculum in established European medical schools. The university lectures and demonstrations fostered a decline in the conventional apprenticeships approach to training. This resulted in a rise of professional standing for the field.

As an upshot, surgeons jockeyed with pure medical physicians for status. Those practicing surgery maintained that their field had its own body of theories and knowledge and was no mere handicraft, but a science. With time the quality of the combined medical and surgical education began to blur the old distinctions between the two fields within the profession.

The best exemplification of this new type of surgeon was Englishman John Hunter (1728-1793). He was a key figure in the transformation of surgery from a simple manual vocation into an experimental scientific discipline involving physiological investigations. Hunter's pupils and other surgeons on the continent produced improvements in a number of operations. A new lateral approach to bladder stone removal, that previously had been a painful and time-consuming procedure, rendered the operation safer and speedier. A French doctor developed a new way to extract a cataract that eliminated the ancient technique of couching that just pushed the clouded lens aside in the eyeball. The invention of the screw tourniquet used in thigh amputation controlled blood flow while the surgeon carried out the procedure was a major advance.

Even with this headway, surgery well into the 19th century was still risky. The surgeon needed to be brisk. As one contemporary observed at the time "the operator's success will be in direct ratio to his quickness." Some surgeons would have their students bring stopwatches to their surgeries. Speed was a matter of pride, but the race wasn't just for show. Every passing second increased the risk of shock, infection, and in the age before anesthesia the bloodcurdling screams.

While swiftness may have been a necessity for the patient, it sometimes was a hazard for those involved with the operation. In one notorious case a well-known surgeon was going so fast to complete an amputation that he also cut off the fingers of his assistant. In those days before sterile surgery, both the patient and the assistant died of gangrene.

Concurrently, with the changes in general surgery, obstetrical skills were progressing. A male surgeon-obstetrician called 'man-midwife' or *accoucheur* displaced the traditional granny midwife figure. He claimed superior skills because of being a qualified medical practitioner armed with an Edinburgh or Paris University degree. His anatomical expertise made him confident that he could leave normal deliveries to nature and his university teachings on how to handle emergencies.

Where the male-midwife surgeons flourished, childbirth was transformed and baby rearing with it. A fashionable lady of the late 18th century might now opt to have her husband present at labor, giving birth in a room into which daylight and fresh air were admitted. The baby was no longer swaddled. And on medical advice this new mother also now breast-fed. Progressive obstetrical surgeons in the late 1700s played a part in changing the theory and practice of childbirth and baby-care.

Hospitals

On a broader community wide level, both in Europe and Colonial America, hospitals in this era were growing in importance for healthcare. They gradually shed their medieval function as a refuge for the poor and became institutions for the care of the sick. But they restricted themselves to non-infectious fairly minor complaints likely to respond to minimal treatment.

Development of the hospitals' primary medical role took time and different ways in various countries. Religious and secular philanthropic movements fueled some institutions while others became the responsibility of the government. Regardless of sponsorship, expansion of existing facilities and new construction quickly occurred.

Many university trained physicians and surgeons migrated to the hospitals to practice and teach medical students. It was the start of clinical rounds. Students were expected to visit patients' bedsides on their own initiative, studying the professors' reports.

But there was downside to the amplification of these institutions. Nosocomial—hospital acquired—infections increased dramatically despite efforts to limit admissions of those with fevers to the hospital. By the end of the 18th century institutionally based disease became so common that hospitals for many became "gateways to death." This would not change substantially until the last half of the 19th century when new scientific theories would be applied to clinical medicine.

CHAPTER FIFTEEN
Medicine in Colonial America

I n Colonial America during the 18th century British officials played only indirect roles in the development of America's formal medicine. The medical initiatives in this time period were like so many accomplishments in this young country—all from within and a product of the uniqueness of the colonists themselves. This adventurous spirit did not result in any worthwhile medical innovations like by the European scientists that embraced the enlightenment's philosophy of the power of reason.

Medical Practitioners

Medicine in the colonies was centered on taking care of the sick. Like in Europe there were several levels of caregivers. Without any American medical schools until near the end of the 18th century, the most elite of the group were physicians that trained at Oxford or Cambridge. Below this class were the barber-surgeons. They were without formal training and assumed an important role in the delivery of care in the colonies.

The third group of medical providers in the colonial period were the apothecaries. They were known as "ye physician's cooke" because they compounded drugs. In small settlements and larger towns, they were more than simply druggists. Because physicians and surgeons were generally unavailable to the lower economic groups, the apothecaries assumed the role of providing medical treatment to the common people.

Apothecaries were often the first to be called upon for medical advice. They mainly relied on medicines that were made from herbs and other plants. For example, chalk was used for heartburn, calamine for skin irritations, and vinegar of roses for headaches. They were a caring and sometimes effective group whose treatments were less harsh than the more traditional therapies like bloodletting.

A fourth unique level of care providers was the minister-physicians. The origin of the clergy adding medicine to their skill set began in England. Here the religious issues between the Anglican and Catholic church saw many ministers concerned that they would be dismissed from the state run religion. Looking for alternative ways to earn a livelihood, some preachers began taking courses on medicine at various universities.

While the first settlements often included doctors, a shortage soon emerged as the colonies rapidly grew and only few medical personnel immigrated to America. In both the Anglican colonies in the south and the Puritan oriented New England, the clergy with their high level of education were asked to deal with both the spiritual and physical ills of their congregations. With access to medical books many clergy became quite proficient as healthcare providers. This was particularly true in New England.

One example was Dr. Samuel Fuller. He not only was a theologian, but also obtained medical knowledge in Europe before coming to America on the Mayflower in 1620. His reputation was enhanced when asked to treat several colonial governors. Another outstanding example was Charles Chauncey who studied theology and medicine in the late 17th century. These and other men extended the domain of minister-physicians and often were the first to adopt new ideas. Overall their function in providing care in those early days cannot be underestimated.

The growth of the clergy becoming doctors may have been influenced by the prevailing viewpoint of many colonists on sin and medicine. This understanding was best expressed in 1718 in Boston by European educated physician William Douglas. He said, "he that sinneth before his maker, let him fall into the hands of a physician. There is more danger from the physician than distemper." Perhaps this clergy physician wanted to reverse that trend for the sinner and hedge his bets on which was more effective, medicine or theology.

The final category of medical practitioners involves women. Most early Americans lived on farms in remote areas. There was little access to a doctor nor could most afford their fees. As a result, the family matriarch assumed the responsibility for curing illnesses. By the 18th century there were books and almanacs available that gave medical anecdotes of various diseases with a suggested treatment. As the preface to one popular volume written by Thomas Dover in 1742 said, "it contained a compendium of disease incident to mankind, described in so plain a manner that any person may know the nature of his own disease."

With this additional knowledge from publications some women expanded their skills outside their immediate household to intervene medically on sick friends and relatives. They became the folk practitioners of the day. These domestic practices did not utilize the harsh treatments of their male counterparts and as a result were often popular. One author commenting on the effectiveness of this approach to medicine noted that "nature does nothing rashly, too much of anything is an enemy of nature."

One area that women provided almost exclusive care in the colonial period was obstetrics. Women midwives followed a long established history of assisting at childbirth. Most midwives had a good record of delivering sound babies. This positive outcome was best achieved when the midwife just patiently waited for nature to complete the birthing process. The best confined their activities to comforting and supporting the mother and finally tying the umbilical cord.

In spite of their skills, midwives in general were looked down upon as ignorant, unskilled, and with unsavory reputations. The worst were those that were too impatient to wait for the natural process and tried various unproven drugs or manipulation to hurry the delivery. At the bottom of the heap were those who deserted women in labor when lured to another patient with the promise of a higher fee. Not surprisingly like in Europe colonial midwives were subject to harsh criticism in the male dominated medical profession of that period. Physicians feared that if women continued to assist in childbirth, they would lose their obstetrical skills.

Diseases in Colonial America:
Smallpox, Yellow Fever and Diphtheria

By the end of the pre-revolutionary years there was a high mortality rate among children and young adults leading to the average life expectancy at birth of only thirty-four years for men and two years longer for women. Those who survived until twenty could expect to live until their mid-fifties.

The particular ills suffered are often difficult to identify in modern terms. Like in 18th century Europe, illnesses in the colonies were characterized by their symptoms that then became the diagnosis. For example, the symptoms of diarrhea and fever were considered diseases. Only the more obvious ailments can be distinguished by symptomatic criteria then available. Those with distinctive signs like mumps, measles and smallpox can be identified, but a general description of "fits, fluxes, and fever" cannot be assigned to a modern name. Moreover, most deaths were attributed to more obscure categories like "old age" or "decay" or the most sinister of them all "being found dead."

The epidemic diseases like smallpox, yellow fever, and diphtheria were the most feared by the American populace. Fortunately, the virulent medieval scourges of leprosy and bubonic plague did not appear in the English colonies. It was the endemic maladies like acute respiratory and intestinal infections along with malaria and consumption (tuberculosis) that were the major killers. Chronic illnesses, malignancy and degenerative disease that afflict present day society were equally fatal in those earlier years. Superficial cancers were recognized, but cardiovascular and kidney disorders were not identified and hidden in such terms as fits, dropsy, and decay. Those illnesses that are more prevalent in the older age groups were less evident and not viewed as major problems.

Most individual maladies were lumped together into a confused mass with little comprehension about the specific characteristics, course, and treatment. The one exception was smallpox. Its striking clinical picture of skin crusts and other symptoms was well understood. Unlike other diseases in the 18th century, prevention with inoculation prevented the scourge of an epidemic.

In the early settlements, smallpox, while devastating to the American Indian population, was not endemic in the colonies as it was in England. It was highly contagious and as known

today is caused by a virus that is transmitted in air from contact with an infected person. The virus could live in clothing
for up to 18 months and be transferred to someone who
shared apparel. Records of colonial towns show that during
epidemics 60% or more of the inhabitants fell sick. There was
a mortality rate of 6-20%.

The isolated scattered communities in the 17th century
provided some protection from the disease. When evident it
did confer immunity to those inflicted. A person with a
pockmarked face was fortunate and was a constant reminder
of the danger of the disease. The lack of a full-blown
epidemic gave rise to an entire generation in the 18th century
without any immunity to the sickness. Many considered
smallpox the principle disease of that century and the only
one with a proven prophylaxis.

The narrative of how smallpox inoculation—not vaccination—
came to America is intriguing. The practice began in old
Constantinople in the 15th century, but by the 17th century it
had spread to Greece. In 1713 a Greek doctor described the
procedure to an English physician, but the idea did not initially
spread. It was not until the wife of the English ambassador to
the Ottoman Empire, Lady Montague, promoted the practice
in Western Europe on her return to England. She had her
children inoculated, as did the Princess of Wales and the Queen
of Denmark. These women's initiative commenced the debate
between physicians for and against the process and their
concerns about the safety and effectiveness of the procedure.

The practice of inoculation or variolation as it was also called spread to America with incredible speed. Lady Montague returned to England in 1721, and inoculations in the colonies began the same year. Initially, it was received with some hostility particularly in the rigid puritanical New England region where interference with nature was frowned upon.

In 1721 and 1722 it took the teaming of a liberal clergyman, Cotton Mather, and a rare physician ally Zabdiel Boylston to start and promote an inoculation drive in Boston. There was a strong campaign against the procedure in pamphlets, newspapers, and from church pulpits. One diatribe against Mather said, "The church ought to deliver him over to Satan to receive the highest censure and scourged out of the country. He should be pilloried and stoned and then banished." An initial prohibition against inoculation was lifted although the clamor against it continued.

In Virginia inoculations were initially illegal. None other than George Washington was an early advocate for the procedure. While he and his brother had been inflicted with the disease and obtained immunity, Washington fighting with the British army before the revolution became aware of the devastating effect small pox had on the troops. In a letter to Patrick Henry, Washington wrote "You will pardon my observation on smallpox, because I know it is more destructive to an army in the natural way than the swords." Later during the hard wintering at Morristown, he ordered soldiers and civilians to be pox-proofed. By 1777, all troops including new recruits were treated.

It is instructive to understand the approved method of inoculation used in America at the time. A healthy donor with a
mild form of the disease supplied clear serum from an early
pustule. The pock was opened with a toothpick, not a lancet
because it might panic the patient. The recipient was prepared
for several days of a mild diet, gentle purges and a bath. The
serum was introduced into an incision of less than an inch in
an area around the deltoid muscle of the upper arm. This was
followed by exposure to outside air, drinking cold water and a
dose of mercurial purgatives. Recovery usually resulted in a
fever and by the twelfth day a pustule formed.

This was a tolerable prophylactic measure that was the first
step in virtually eliminating smallpox worldwide. The dangers
of variolation were put aside in 1798 when English doctor
Edward Jenner (1749-1823) electrified the world with his
famous paper on cowpox and the benefits of vaccinating
humans with the fluid from the sores of "vaccinia," a disease
of cattle.

Jenner learned about this from the country folks in his native
Gloucestershire. Cowpox was a known disease occasionally
contracted by dairymaids. The young Doctor Jenner observed
that when this occurred it conferred immunity against smallpox.
In an experiment to prove his hypothesis, he inoculated an
eight-year-old boy with debris from a dairymaid's cowpox
pustule. The boy developed a slight fever from which he soon
recovered. Six weeks later Jenner then inoculated him with
smallpox "virus." The inoculation did not take, proving that the
cowpox vaccination worked. Further observations demonstrated

that this new vaccination had none of the potential dangers of the earlier inoculation of material from smallpox itself.

Contrary to the stormy reception given smallpox variolation 60 years earlier, the introduction of vaccination in America created little furor. It was hailed by leading American and European physicians as the most useful discovery of the 18th century. Today, smallpox is the first disease to be completely eradicated in the world. The last case in the United States occurred in 1949, and routine childhood vaccinations ended in 1972.

The second most dreaded epidemic disease of the colonial period was yellow fever. It was known as "The Great Sickness," "The American Plague" and "Barbados Distemper." While it did not have the mortality rate of smallpox, it struck with such ferocity in the port cities where it spread gloom and fear throughout the region.

Yellow fever was considered a disease of the tropics. Typical outbreaks began with the arrival of a ship that had previously stopped in the West Indies. The first appearance in Colonial America was in Philadelphia in 1668. The same city was affected in 1699, 1741, 1762 and the worst time was in 1793 when it was the nation's capital. There were reports of epidemics in New York and Boston in 1693, in Charleston in 1745, and in Texas and New Hampshire. People had to work day and night just to bury the dead.

Today, it is known that the Aedes aegypt mosquito spread the yellow fever virus. This mosquito's most distinguishing feature was its preference for the urban habitat around standing water and cisterns. The larva brought from the West Indies in the cargo ships flourished in this environment. In the colonial period where homes had no window screens, these flying insects had ready access to life inside a house. As the terrifying disease broke out it spread from person to person by the tainted mosquitoes. Unlike the endemic malaria mosquitoes, the yellow fever virus carrying mosquitoes did not survive the winters and had to be reintroduced after the frost season.

The symptoms began with a headache, backache, and fever. People became very ill from the start. It is a form of hepatitis that turned the skin a yellow color in the third day of the illness, thus its name. As the disease matured so did the horror of the symptoms. There could be bleeding from all orifices including the pores of the skin. Vomit came out black. At the end of a week the afflicted person was either dead or recovering. For those that survived, one exposure to yellow fever conferred a lifetime of immunity—a fact not understood at the time.

Epidemiologically it is different from most infectious diseases that are prone to strike the elderly and children. Yellow fever was more likely to occur in males between 15 and 40. While most other contagions spread among the filth of the poor, the yellow fever mosquito could thrive in the houses of the rich and powerful.

Benjamin Rush was one of the first to recognize the disease as yellow fever. He took an activist approach to treatment.

Rush believed that copious bleeding was necessary. He wrote that "the physician must not falter and deny treatment just because the patient is weak." His advocacy of bleeding and purging even when there is a feeble pulse and bloody bowel movements promoted a feud with other doctors during the 1793 epidemic. Many refused to follow his treatment recommendations. Rush paid for a newspaper advertisement that addressed the people of Philadelphia criticizing doctors who spoke against his therapy.

Some folk healers recommended using vinegar as a prophylaxis to the disease. A hot iron in a bowl produce fumes that may have discouraged the mosquitoes—although it was not known at the time that they were the vectors of the disease. Some caregivers thought a tarred rope in the hand and a camphor bag necklace produced immunity. Others chewed garlic throughout the day to keep the contagion away. None of these practices proved effective.

An historic time line delineates other outbreaks of yellow fever extending into the 19th century. Its cause was discovered in 1901 by Walter Reed. The public health preventive measures of the disease made the construction of the Panama Canal possible. A vaccine was first discovered in 1936. All this was too late for the folks in Colonial American times.

The third epidemic infectious disease was the saddest. It was diphtheria that was also known as the "strangling angel of children." In the colonial era it was lumped with scarlet fever

although as now understood they were of different bacterial
etiologies.

Descriptions of diphtheria-like illnesses first appeared in ancient Egypt and Greece. In early America some recorded
"throat distemper" in 1659. But it was the 1735-1740 outbreak in New England that was one of the most dismaying
epidemics in the prerevolutionary years. It targeted children
and ill-fated families in rural New Hampshire and
Massachusetts. The communities involved were isolated,
widely dispersed and highly susceptible. In some of the towns
nearly 50% of the children died. Parents grieved in the village
meeting houses and prayed over the bodies of the little one
who succumbed so quickly to this deadly disease.

Some of the cases may have been either scarlet fever or
diphtheria. Without bacterial evidence, it is difficult to
determine retrospectively whether the contagious bacteria
corynebacterium diphtheriae or streptococcus bacteria of scarlet
fever caused all the cases. At that time the origin was thought
to be from an unknown source circulating in the air. Scientists later found that the disease is mainly transmitted from
person to person directly by droplet infection. It can also
spread in clothing, bedding and in untreated milk.

The presenting symptoms of the two entities were clinically
similar. Unlike diphtheria, scarlet fever had long term effects
on its victim. But it did not have the same degree of immediate mortality. In the case of diphtheria, the symptoms included sore throat, fever, difficulty swallowing caused by the

bacteria entering the mouth and settling on the back of the throat. As the disease worsened swelling of the neck caused difficulty breathing. The lymph nodes enlarged giving a "bull neck" appearance. In severe cases death from suffocation occurred. The most dangerous result was when the toxin entered the blood and produced heart failure.

The colonial era physician's inability to manage the disease except with the devastating bleeding of the ill children should be of no surprise. Painful incisions were made under the tongue of the sick child to initiate the bloodletting. It had no curative effect.

During the decade of 1850-1860 a worldwide pandemic of diphtheria broke out. In the 19th century tracheotomies where used to bypass the swelling in the throat. When the bacterium was identified in 1883, an antitoxin was developed. In the 1930s a vaccination became available for the disease. While it is still seen in some poor countries, it has been eliminated in the developed world.

The infectious diseases of small pox, yellow fever and diphtheria had a special visibility and cultural salience in the history of colonial America. They were episodic, unpredictable, and frightening. They were a highly visible component in the collective experience of these early citizens. Unlike epidemic maladies, endemic diseases killed and disabled on a regular basis. They were unavoidable and part of the conditions of colonial life. The most quintessential was malaria.

Malaria was one of a number of different fever causing sicknesses like typhoid and dysentery that inflicted the early settlers. It was the most common long run offender. The first known case of malaria in America was after Columbus arrived in 1492. For the English settlers it was the Africans who came to America as slaves that started the illnesses in earnest. In addition, some original founders and indentured servants that traveled to the new world from England and parts of Europe carried the malaria parasite with them.

From the English settlements in Virginia, the disease moved south to the Carolinas and beyond. The spread was hastened by the stagnant shallow water needed in rice cultivation. Although not known at the time, this environment was very welcoming to the mosquitoes carrying the disease. From the southern states it gradually traveled north to Maryland and westward to Ohio. Outbreaks of Malaria occurred as far north as Massachusetts in the 17th century and elsewhere along the North American coast throughout the colonial period.

Malaria is caused by the protozoan parasite Plasmodium. Mosquitoes carry the organism and spread the disease from person to person. The symptoms usually begin 10 days after the mosquito bites. The parasite enters the human body and the person's red blood cells. Those inflicted have fever, headache, and vomiting. The majority of the early settlers became debilitated and some died from the infection—newcomers had a higher mortality rate. Typically, most infected individuals would appear normal after several days. Once acquired it naturally reoccurred within the infected individual. Malaria became known as "intermittent fever." Because of the abundance of mosquitoes, the infection was assured to effect new colonists.

In the 17th century there were no treatments in colonial America. But in the 16th century in South America the Peruvian Indians introduced the bark of the cinchona tree to the Spanish explorers to alleviate malaria symptoms. This was a genuine specific treatment against malaria—it is known today that the bark contains quinine. It was not until 1720 that it reached the English colonies via Europe. It was one of the few effective drugs in colonial America. It was used by the practitioners to treat all kinds of fevers because they did not recognize its specificity for malaria. It was the aspirin of that period. As standards of living rose during the 18th century, the mortality from malaria fell as communities matured and living conditions became more comfortable. The scourge continues today with various virulent subtypes that are usually resistant to quinine. Treatment has improved and potential vaccines are in the works.

Hospitals

By the end of the colonial period in 1783 there were only two active general hospitals: Pennsylvania Hospital in Philadelphia and Charity Hospital in New Orleans. For most patients these facilities were a last resort. Initially, colonial hospitals were no more than shelters for those who could not afford treatment at home. This aligns with the origin of the word "hospital" which comes from the Latin for "immigrant home or visitors' home."

There were hospitals in the New World before the British came with their permanent colonies. The French and Spanish were much more proactive and advanced than were the English. The Spanish established hospitals early on in their conquests. In Mexico alone within the first 100 years during the 16th century 125 installations were built including many for the Indian populations in central America. The peripheral Spanish settlements in the area around St. Augustine, Florida had less formal organized facilities. After 1697, there were a few small thatched structures periodically erected. There were similar provisions present in the Spanish outposts in the southwest.

The French took longer to establish hospitals in their settlements. Two modest institutions were founded in Quebec City in the 1640s. By the end of the 17th century three more had been erected. All of them were staffed by nuns from various Roman Catholic orders.

In the English colonies the first hospital was built in 1612 along the James River in Virginia. It was 80 beds in size and was lodging "for the sick and lame, with keepers to attend them for their comfort and recovery." It burned to the ground in 1622 during an Indian uprising and was never rebuilt. Probably the only other hospital of the 17th century was a small military structure erected in New Amsterdam in

1658. In the 18th century the first "real" hospitals were established.

There is historic controversy between Charity Hospital in New Orleans and the Pennsylvania Hospital in Philadelphia as to which one should be considered the oldest existing hospital in the United States. Part of the befuddlement revolves around the definition of a hospital and almshouse. In large towns during the 18th century almshouses were established to take care of the chronically ill and aged poor. Some of these institutions provided medical and surgical care for acute sickness and injury.

In 1752 the Pennsylvania Hospital was founded through the efforts of Benjamin Franklin and Dr. Thomas Bond. Unlike almshouses that admitted all patients, anyone with a disease judged to be incurable was not allowed in this institution from the beginning. Meanwhile, Charity Hospital was known to be established in 1726, a full 16 years before Pennsylvania Hospital opened its doors. But the historic supporters of Pennsylvania Hospital say that Charity in the early years was an almshouse and not a true hospital. As a result, Pennsylvania Hospital claims to be the oldest true hospital in the United States. With Charity Hospital's destruction by hurricane Katrina in 2005, Pennsylvania Hospital became the oldest continuously operating institution.

While the Pennsylvania and Charity Hospitals may have been the first general care institution in the colonial period, there were other smaller facilities erected for special groups. These

included inoculation hospitals or pest houses and a few separate institutions for the mentally ill. The pesthouses were
precursors to the modern hospital. They were designed to
protect the public from infectious diseases by isolating patients considered at the time to be contagious. In 1702
Boston citizens with smallpox were sent to a pesthouse. In
1738 a special installation was established on New York har-
bor's Bedloe Island to isolate infectious persons arriving by
sea thought to have a communicable infirmity.

Pennsylvania Hospital was a success. It was founded with
some provisional funds and was governed by a private self-
perpetuating board. In its first year it admitted 64 patients
with over half listed as cured. Unlike other hospitals, if after
providing for the poor, there were rooms to spare private patients could then be admitted. They would be charged "at
such reasonable rates as they can agree to." The presence of
private patients may have maintained upper-class interest and
promoted higher standards than in a strictly charitable institution. While caring for private patients was secondary to the
main goal of providing care for the sick poor, the admission
of paying patients was an historic first. The success of the
Pennsylvania Hospital model encouraged those in New York
to build a similar institution in 1791.

The establishment of the Pennsylvania Hospital was one of
several changes that occurred in American medicine in the
mid 18th century. While there was limited progress in the
colonies, the greatest amount of scientific and medical
knowledge remained in Europe. Here professional societies

were formed in all major European capitals, and scientists shared their research in published journals.

First American Medical School

The European scientific community made medicine more exciting. This stimulated medical teaching and the standards of medical education rose particularly in Leyden, Edinburgh, and London. It was to these centers that beginning in about 1730 Americans of means migrated to formally learn medicine.

The medical school in Edinburgh was the mecca for American students. It was the premier place to study. Traditional medical teachings were no longer followed. Admission standards were liberalized but required a prerequisite of broad knowledge in a variety of fields not just Latin and Greek. Courses included anatomy, surgery, chemistry, pharmacy and clinical lectures in the associated Royal Infirmary. It took nearly 4 years to receive a degree in medicine.

The American students in Edinburgh were hard working and diligent in their learning. They grouped together for study and recreation. In their free hours they often compared the university medicine to which they were exposed to the low standards they remembered in America. It was within this community that John Morgan and William Shippen planned how they would start a new medical school on their return to America based on the Edinburgh model.

Dr. John Morgan ultimately emerged as the initiator of the first medical school in the 13 colonies. One biographer went further and labeled him "the founder of American medicine" and "medicine's main attraction." He was a handsome dandy who wore the latest in colonial attire and scandalously twirled a silk parasol as he walked the streets in Philadelphia.

Morgan persuaded the College of Philadelphia—later the University of Pennsylvania—to establish a medical faculty in 1765. He introduced the Continental-Scottish tradition of a university based medical education. With his background of hospital training in England he also emphasized the need for bedside teaching. This took place at the Pennsylvania Hospital. The original faculty consisted only of Morgan and Shippen but others were added including Benjamin Rush who taught chemistry.

The school was an immediate success. At the first commencement 10 students received their medical degree. With this initial class Morgan's vision of having graduates raise the professional standards of medicine throughout the colonies began. It emphasized superior learning and practicing only as physicians. In other words, they would not do surgery nor sell drugs. This was poorly accepted at the time because most apprentice trained general practitioners of the day did a little of both. But Morgan's model was actually prophetic. Today, despite a common medical school background, the practice of surgery has become a distinctive specialty, and pharmacy is an independent profession.

Morgan's program was a comprehensive template that went beyond just beginning a new medical school that introduced superior British practices into the colonies. Few of his ideas were new, but they were considered progressive in nature in the conservative American setting. One novel and controversial part of his plan was an effort to have formally educated physicians control the practice of all medicine to the exclusion of those less qualified. Morgan wanted physicians to create a united and prosperous profession. He proposed to accomplish this by forming an elite group of doctors that would license all practicing physicians. Despite his efforts the overwhelming majority of colonial doctors continued to practice a combination of medicine, surgery, pharmacy, and midwifery without undergoing a licensing examination.

Morgan was an advocate for more medical research. This met resistance because of the popular aversion to human dissection. Without this procedure expansion of pathological anatomy was impossible. The unreasonable public attitude about autopsies was manifest in the notorious "Doctor's Mob" in New York City. Here an unruly crowd in 1788 attempted to lynch those medical students engaged in autopsies.

While there was opposition to some of Morgan's ideas, his strategy of a university-affiliated medical school augmented with hospital-based instruction quickly became accepted as the blueprint for the future development of American medical education. The medical school in Philadelphia was soon followed by one in New York in 1768 at King's College. This later became Columbia University. Medical institutions in

Boston came somewhat later. The first in 1783 was associated
with Harvard.

Morgan's proposition to govern the practice of medicine was
not unique. It had been around for many years. In medieval
England the licensing of doctors was under the church's
jurisdiction. During the Renaissance the universities and the
professional guilds regulated the doctors. In Colonial America
some tentative steps were taken, but they were not effective.

A short lived 1736 Virginia law set standard charges for doctors. This regulation provided university trained physicians
with twice the fees of the apprentice-trained doctors. The
most well thought out canon in the colonies was in New
York in 1760. It required applicants for a medical license to
be examined by government officials and reputable physicians. Despite it soundness it was not enforced. No one was
ever convicted of practicing medicine without a license. A
statute was passed in New Jersey 12 years later in 1772 with
similar results. With the advent of the American revolution,
medical licensing directives receded.

Benjamin Franklin and Benjamin Rush

Two well know men had a significant influence upon medicine
and in the larger sphere many other aspects of Colonial
America. They were the "Benjamins" of the revolutionary
generation—Benjamin Franklin and Benjamin Rush. They
shared the status as enlightened Renaissance men and
Philadelphians.

Franklin played varied roles in his long extraordinary life including: printer-publisher, civic activist, revolutionary, statesman, scientist, philosopher, diplomat and sage. Is it any wonder that with all this activity his medical interest attracted less attention? But, in fact, he had a great deal of interest in health related subjects. He was a medical activist and inventor.

Franklin applied Enlightenment reasoning to various afflictions. He was frequently very accurate in his theories. He was one of the first to discover the health risks of exposure to lead by observing the effect it had on laborers whose trades used the metal. While in France, he noticed that many individuals with abdominal pain came from workplaces that utilized large amounts of lead. He reported his conclusions to local French doctors. They subsequently found lead related health risks in other tradesmen and also in lead contaminated food.

Franklin promoted a moderate diet, exercise, and self-control in all things. He noticed the effect of a vigorous workout on his heart rate. He advocated a healthy lifestyle in his *Poor Richard's Almanac* with his well know phrases:

- "Early to bed early to rise, makes a man healthy, wealth, and wise."
- "Be not sick too late, or well too soon."
- "Time is an herb that cures all diseases."
- "Eat to live and not live to eat."

Before viruses and bacteria were discovered Franklin perceived that the common cold was passed between people through the air. In the 1700s most of the populous and practitioners believed that wet clothing and dampness in the air caused a cold. Franklin deduced that while sailors' clothing was always wet they were generally a very healthy bunch. He concluded people caught colds from one another in small rooms sitting close together and breathing on each other—quite an inference for that early era.

He produced medical inventions including the flexible catheter. It was designed for his brother who suffered from bladder stones and urinary retention. It was made of silver wire, coiled with joints to allow flexibility, and covered with gut. His most famous innovation was bifocals or "double spectacles." As he aged, Franklin found it awkward shifting back and forth between regular and reading glasses. He cut the two lenses in half and joined them in one frame. Bifocals have served millions of people since he invented them in the 1750s.

While not formally trained as a physician, Doctor Franklin—he was called this after receiving an honorary degree in 1759—was a role model for physicians then and now. His capacity for detailed perceptive insights, fastidiousness in recording observations, analysis of scientific phenomena, and knowledge of human conduct combined to improve the lot of his fellow man.

Benjamin Franklin enjoyed greater fame than Benjamin Rush who was better educated, more widely traveled and more broadly experienced. Franklin's mystique came from his status as a self-made man without formal education. By contrast, Rush was schooled on both sides of the Atlantic. He did his undergraduate work at the College of New Jersey—the early name for Princeton—and began his studies in Philadelphia before receiving his medical degree from the University of Edinburgh. After graduating he spent several years in Europe practicing medicine while studying French, Italian, Spanish and science. He returned to Philadelphia in 1769 and opened a private practice. He was appointed professor of chemistry at America's first medical school where his classes were popular.

Medically Rush was of the old school. Like many he argued that an imbalance in the body caused malfunctions physically and mentally. He was committed to treatments that considered bleeding and active purging to be the cure for nearly every human ailment. During the yellow fever epidemic of 1793, he gained acclaim for remaining in town and treating some 100 patients a day with vigorous bleeding. Many of his fellow doctors vociferously objected to Rush's extreme use of bloodletting. But he persisted with the treatment throughout his years in practice even insisting upon being bled himself shortly before his death.

Rush's most accomplished and long lasting medical achievement was in the diagnosis of mental illness. He was considered the "Father of American Psychiatry" publishing the first textbook on the subject in America. Rush classified different forms of mental illness and theorized as to their causes and possible cures. He was one of the first to describe

'Savant Syndrome' in those with exceptional abilities in an isolated area. He also pioneered the therapeutic approach to alcoholic addiction believing it was a medical disease.

Outside of medicine Rush was best known as a leader in the emerging nation. As a member of the Continental Congress he signed the Declaration of Independence. During the Revolutionary War in his position as Surgeon General he lost some of his luster for his criticism of the army's medical service. He complained to George Washington who deferred Rush's concerns to Congress. Rush resigned after his position was not upheld. After the war he advocated the adoption of the Federal constitution and was appointed treasurer of the United States Mint—a position he held from 1797 to his death in 1813. When he died he was eulogized as the "American Hippocrates."

Medicine During the Revolutionary War

The Colonial period of our nation comes to end with the American Revolution. The fight for freedom and separation from England was costly in the lives that were lost by those engaged in the conflict. No reliable statistics are available on the causality rates. Estimates for the number who died from 1775-1781 are between 30,000 and 70,000. More telling is the nine to one ratio of deaths from sickness to battle causalities.

The military medical phase of the struggle has been characterized as chaotic, chronically discouraging, and sometimes calamitous to the point of threatening the whole war effort. There were approximately 1,400 physicians and surgeons enrolled in the American armies during the Revolution—some were present for the entire war while others for a single campaign or battle.

How physicians and surgeons integrated into the troops to provide medical care is a confusing scenario. It involved strong personalities, organizational conflicts, and the lack of effective therapy. The overall picture was further muddled because neither a strong central government nor a national army was present at the outset of the revolution. There was a desperate effort on the part of the breakaway colonists to create a large medical organization from nothing and keep it running with limited resources throughout the fight.

The medical care of the troops depended upon a number of interacting factors including:

(1) The quality of the surgeons and hospitals
(2) The availability of food, clothing and medical supplies
(3) The willingness of Congress or the provincial legislatures to provide sufficient appropriations.

In the beginning, the army's military commanders as well as the national and provincial legislators grossly underestimated the requirements to care for the sick and wounded soldiers. Even General George Washington did not immediately realize the need for more extensive medical services. As late as January 1777 he marched his army from Trenton to

Princeton without taking a single surgeon or communicating to his medical officers of the movement.

Part of the organizational problem was in the different directors that led the army's medical department during the conflict. Each had unique styles and most had confrontational personalities reducing their effectiveness. The first, Dr. Benjamin Church, was accused of infringing upon regimental surgeons' authorities when he began consolidating various hospitals. While criticism of Church's work mounted , he was discovered to be in a treasonous correspondence with the British. He was dismissed and allowed to leave for the West Indies because of failing health. The ship he was on for the journey was lost at sea.

The second director was the well-known Dr. John Morgan, the founder of America's first medical school. He was selected by Congress in October 1775. While well qualified and energetic, his strong personality eventually was his downfall. His immediate problem was to obtain medical provisions. In doing so he created dissension with some regimental surgeons who accused him of hoarding supplies. In fact, it was Congress that had neglected to fund for provisions.

On top of this allegation, he led a campaign to ensure that the regimental surgeons were better qualified. Morgan described them as "unlettered, ignorant, and rude to a degree scarcely imaginable." In the beginning of the hostilities colonial American practitioners had to be transformed from common medical providers into military surgeons. With no formal instruction to master the surgical skills needed in warfare, most doctors depended upon the few surgical texts available in America.

While Morgan sought to improve the caliber of the surgeons, it did little to enhance his reputation. Congress undermined his authority by appointing others beneath him without defining their relationship to Morgan. The disharmony that ensued led to his dismissal in January 1777.

Much to Morgan's surprise and outrage his old rival William Shippen was named as his replacement. Shippen was more polished and was able to work with his assistants who were named by Congress to reorganize the military hospitals. Regional institutions were established with an appropriate chain of command. This helped, but it failed to stem the discord between the various regimental surgeons.

Throughout his tenure Shippen was constantly being out maneuvered by his underlings with political influence. They bypassed Shippen and took their problems directly to Congress that then tried to intervene in departmental affairs. There was a famous case of interference between Shippen and Benjamin Rush who was one of the assistant surgeon generals. The vindictive attacks between the two headstrong doctors eventually led to Rush's forced resignation from the army in January 1778. Not to be outdone, Rush then accused Shippen of profiteering on hospital supplies. A long legal battle ensued that included a court-martial where Shippen was found innocent. Behind the scenes Rush continued his attack.

He had Congress review the court's decision. After some time without any Congressional action, it gave Shippen some measure of face and he resigned in January 1781.

The final wartime medical director, Dr. John Cochran, was not only very capable but was able to avoid personality clashes that had troubled the department since its inception. The military action ended in 1781, but Cochran remained on the job until the peace treaty in Paris was signed in September 1783. By that time the majority of the sick and wounded had been discharged.

In all the confusion in the leadership of the army medical division, it is easy to forget about what happened to the sick and wounded during the fight. It is not surprising that American causalities were greater percentage wise than the British. This was likely because the English troops were veterans and already had survived many of the main encampment health disorders. These maladies had a devastating effect upon the colonist recruits. In addition, the British were better disciplined and had a more effective medical service.

On the American side the major disorders affecting the troops were smallpox, dysentery, respiratory complaints, malaria, and camp fever—likely typhus and typhoid. Malaria was an issue for soldiers in all regions but more so in the southern colonies. During the siege of Yorktown, the New England soldiers in particular suffered from malaria.

Unfortunately, the day-to-day accounts of the military surgeons and doctors was never recorded in detail. One
Frenchman traveling through North America during the war
observed: "I make use of the English word doctors, because
the distinction of surgeon and physician is as little known in
the Army of George Washington as in the time of
Agamemnon the Greek." Surgery was not even attempted for
major chest or abdominal wounds. An injured soldier usually
meant hemorrhage from the injury. Compression was used
for minor bleeding, and ligatures for major trauma. For gun-
shots the ball was removed when possible, but infection
usually followed.

The injured and sick were sent to hospitals, but most of the ill
colonist soldiers simply went home to be treated. Large
hospitals in the major cities were evident, but as the war
accelerated deaths in the large institutions skyrocketed and
small hospitals came in favor. Like for civilians the many
drugs that were used to treat the troops were of little value
except for so called "capital drugs": Peruvian bark for malaria
and undefined fevers, opium for pain, and calomel-mercury
for venereal diseases.

By the end of the war the nation had achieved independence.
From the standpoint of medicine and surgery no significant
advances were made during the conflict. Even with the primitive knowledge of the day much of the carnage occurred
needlessly because the army medical department lacked an

efficient chain of command. Those that were in charge where more often concerned with their own self-promotion than advancing better health and care for the troops. This attitude was reinforced by those in the Continental Congress where political influence and infighting made them fail to understand the enormity of the problem.

The America colonies separation from England brought little change to the number of providers serving the populace. Since the British had few structural or formal medical presence in the colonies, there were no medical personnel that had to be replaced with their departure. The outbreak of war did affect those American medical students in Edinburgh and London. They had to leave their classes and return home. Those physicians that were Tories fled to Canada and England. As the war ended those patriot physicians that had been in the army returned home to a serious shortage of medical supplies brought on by the British blockade. The civilian populations suffered, but the troops had suffered more.

Post-revolutionary war medicine would not advance to a higher level with the same extraordinary speed that the severing of ties with Britain had on the political, social and economic parts of the nation. Medicine in Colonial America was primitive in the beginning and remained so throughout its existence. While progress was made in inoculation and medical education, the basic medical theories were still elementary. Medical advancement in America would not begin until future scientific theories and discoveries were applied to the practice of the average physician.

For many historians the formation of the United States at the conclusion of the American Revolution is the end point of the Enlightenment era. For medicine particularly in Europe there was a gradual change in values and perception regarding health as a positive tenet. It was a period of consolidation and systemization of medical theories.

Faiths in medical science contributed to a new confidence in man's capacity to master his environment and challenge religion's meaning of life. Medicine was slowly gaining cultural authority. This positive effect was offset by the conservative approach of physicians of the period on the continent and America. They were unwilling to absorb and utilize some discoveries of their more advanced thinking colleagues. At its best, the record of Enlightenment medicine was mixed and set the stage for the imminent gains that would occur in the 19th century. It was then that medicine became truly scientific.

CHAPTER SIXTEEN
The 19th Century
The Dawn of Modern Times

P rior to 1800 medicine was as much an art as science. Modern times commenced with the 19th century, and healthcare became scientific. The medical profession freed itself from the morass of old theories and systems. Doctors made a sweeping return to clinical observations with the advent of new technology and discoveries.

Modern medicine is built upon the foundations that were established between 1800 and the onset of World War I. During this time there were parallel developments in the bio-medical sciences of physiology, pathology, microbiology, bacteriology, immunity and clinical practice. While this was occurring, everyday life beyond healthcare was altered by innovations: steam ships, railroads, power flight in 1903, electricity, telegraph, telephone and the motor car to name a few.

New scientific organizations were founded, and reformers declared science the powerhouse of progress. Directly and indirectly, medicine benefited from such transformations. Medical practices were slowly modified into a modern appearance that still characterized the profession today even with the vast modifications that further innovations have wrought.

The most remarkable discoveries happened in the latter half of the 19th century following the introduction of the germ theory by Louis Pasteur. But there were also a number of achievements in the early 1800s. These include the introduction of anesthesia in surgery, advancement of cellular pathology, the invention of the stethoscope, and understanding the digestive process.

Anesthesia Changes Surgery

During the 18th century there was some progress in the technical aspects of surgery. But from the patient's perspective, pain was a powerful reason for avoiding even the most desperately needed operation. While narcotics had been around for centuries, they were rarely used for the relief of surgical pain. Alcohol was tried, but the large doses necessary to induce stupefaction caused nausea, vomiting and even death instead of sleep.

Sometimes a direct but crude way of inducing a state of insensitivity was just to knock the patient unconscious with a blow to the jaw. This method was not very specific or complete. The surgeon might be able to extract a bullet before the patient recovered from the shock of the left hook. Those choosing not to be punched could select to drink hemlock. It was sometimes used in anesthetic concoctions with mixed results.

British theologian, political theorist and chemist Joseph Priestley (1733-1804), who was best known as the discoverer of oxygen, came up with the gas known as nitrous oxide or "laughing gas" in 1772. Initially, charlatans promised miraculous cures for all sorts of disorders from inhaling this vapor. It was snorted through the upturned noses of idling English gentry at parties. Eventually some tried to find legitimate uses for this new gas.

The breakthrough came when the bon vivant English scientist Humphry Davy (1778-1829) began experimenting with nitrous oxide to see if it would soothe a hangover. One day he suffered from a severe toothache, and he inhaled the gas for relief. In addition to feeling giddy, relaxed, and cheerful Davy noted the discomfort caused by his infected tooth had almost disappeared. Recognizing that the gas seemed capable of destroying pain he speculated that it might be used with advantage during surgical operations. Unfortunately he did not pursue this line of reasoning. So his great idea rested on the back shelf for years. Decades of surgical patients went without pain relief.

The next decisive step took place in the United States four decades later. In Connecticut dentist Horace Wells (1815-1848) in 1844 successfully started the practice of anesthetizing his patients with laughing gas. At the time dentists were more highly motivated than other health practitioners to discover novel and powerful anesthetics. This was because until the excruciating pain of a rotten tooth exceeded the anticipated agony of extraction, the victim of a toothache was unlikely to submit to the services of a dentist.

Meanwhile, ether had been around for a long time. The honor of being the first to synthesize ether has been attributed to several different Renaissance alchemists. Like nitrous oxide it was used as frolic for years without anyone recognizing its anesthetic properties. It took American William Thomas Morton (1819-1868) to pioneer ether. Much to the chagrin of many, he was an unlicensed dentist and fast talking con man that borrowed or stole the idea of using ether as an anesthetic.

He developed a new inhaler and added color to the ether so that he could then falsely labeled it as a patentable new gas. He called it "Letheon." He hoped to benefit financially by expanding its use beyond dentistry. Morton approached a Boston surgeon who agreed to try it. This famous trial operation took place at the Massachusetts General Hospital on October 16, 1846. It was a full success medically. But Morton's charade was uncovered, and he was disgraced and lost his credibility. The operation would end the era of the surgeon being portrayed as an "armed savage" and eventually allow a remarkable series of improvements in the field.

The triumphant of the operation in Boston using ether was widely reported. Ether quickly spread to Europe. Inspired by its successful use, Scottish surgeon and obstetrician James Young Simpson (1811-1870) in 1847 initiated a search for another anesthetic that did not have ether's disadvantages of being an explosive, irritable to the lungs, or cause vomiting.

Using himself and colleagues as guinea pigs he sniffed his way through multiple known reagents until he finally came upon chloroform. The whimsical story about its discovery was that Simpson and his assistants were testing chemicals at his house when somebody accidentally upset a bottle of chloroform. Shortly thereafter on bringing in dinner for the group, Simpson's wife found them all sound asleep.

Simpson then tried chloroform on a woman in labor He was so pleased with the result that he administered it to other pregnant women to reduce the discomfort of the birthing process. Clergyman and others denounced Simpson proclaiming that it was unnatural to alleviate the pains of childbirth. They felt women should endure it with patience and fortitude. This opposition quietly disappeared when on April 7, 1853 Queen Victoria took chloroform for the birth of Prince Leopold. Another more recent member of the royal family, the late Diana, Princes of Wales, may have said it best about the uncomfortable nature of childbirth when she remarked that "if men had to have babies, they would only ever have one each."

General anesthesia as well as the introduction of cocaine as a local anesthetic in the late 19th century proved invaluable for deadening pain. But these anesthetics did nothing to address the root cause of another serious problem hindering surgical advances—infection. A solution to this would come later in the century.

Rudolf Virchow—Father of Microscopic Pathology

While the discovery of anesthesia had wide acceptance in practice, German Rudolf Virchow's (1821-1902) cell theory radically altered the direction of medical thinking for the future, but it had no immediate application for the patients of the period.

During the 19th century Virchow was the best known and most respected medical man of his time. He was called the "pope" of medicine because of the eminence of his scientific influence. Virchow symbolizes the shift from gross to microscopic anatomy in the pathological field that reigns supreme to this day. It is significant and in keeping with the spirit of the times that his leadership spearheaded the increased importance of the laboratory medicine.

In 1858, Virchow's book, *Cellular Pathology,* presented his groundbreaking theory. Proven by microscopic observations, he established that the cell is the ultimate unit of pathological disturbances and of normal life. This is summarized by the phrase "all cells from cells."

Virchow postulated that it was theoretically possible to trace every disease back to its primary locus. Cells were the principal location of pathology because they were also the units of normal functions. Virchow believed that the cell provided a unifying focus for both physiology and pathology. It demonstrated common features of inflammation, pus formation, and tuberculosis growth. In his cell analysis on tumors he brilliantly proved that cancer arose from abnormal changes within cells that then multiplied out of control through metastasis. In the process he was the first to describe different kinds of leukemia.

Throughout his long life his indefatigable energy and voracious mind explored a variety of other fields besides pathology including: anthropology, archaeology, history, politics, public health and sociology. He was indeed the complete man of medicine. Virchow probably never suspected that because of him microscopic pathology would become one of the foundations of medicine a century later.

Physical Diagnosis Advances

A synthesis between anatomy and clinical medicine took place in the early 19th century with the advent of improved physical diagnosis. Up until then methods of examination before death revealed little about internal organs. Physicians believed that to have a disease a person had to feel sick. This all changed with an enhanced approach to bedside examinations that incorporated the invention of thoracic percussion by Auenbrugger of Vienna and auscultation via Laennec's stethoscope in Paris. Symptoms of living patients could now be linked to anatomical changes. The role of the cursory physical exam of the patient that had been present for centuries would be changed forever.

In 1761 in his *Inventum Novum (New Discovery)* Austrian physician Leopold Auenbrugger (1722-1809) announced the technique of percussion of the chest. An innkeeper's son, he had been familiar with the trick of thumping barrels to test for their fullness. Moving from kegs to rib cages after medical school, he recognized that when struck with the finger, a healthy chest sounded like a cloth covered drum. By contrast, a muffled sound or one of high pitch indicated the patient had pulmonary disease. Auenbrugger's colleagues initially ignored this new method of physical diagnosis.

A young physician in Paris, Jean Nicolas Corvisart (1755-1821) , eventually picked up on the technique. After over twenty years of experimentation he published his translation and revision of Auenbrugger's treatise. Students flocked to Corvisart's clinical rounds and watched him predict anatomical findings with surprising success. Percussion soon became a part of the process of evaluating the ill patient.

Another Paris doctor, Rene-Theophile-Hyacinthe-Laennec (1781-1826) would take physical diagnosis to an even higher level by devising the stethoscope in 1819. It became the chief tool for objective signs of disease until the discovery of x-rays 77 years later. The stethoscope changed the approach to internal disease and hence the doctor-patient relationship. At last, the body was no longer a closed book. Pathology could now be detected on the living.

As a young child Laennec's mother died. Soon afterwards his father abandoned him to the care of his brother who was a physician. The uncle's influence led the young Laennec to Paris where he was an academic success in medical school. The invention that brought Laennec fame came about because of the sensitivity at the time of a doctor actually touching a patient during their physical evaluation.

In examining a young stout female with a suspected heart problem, Laennec needed to listen to her heartbeat. He was too embarrassed to actually touch his ear to her skin. So, he rolled sheets of paper into a cylinder and touched one end to her heaving chest. The other end he put to his ear and listened. The sounds of the heart and lungs were magnified. Of equal importance, he preserved both his dignity and her privacy. He named his invention the stethoscope from the Greek words for chest and to explore.

Over the succeeding years Laennec constructed a more stable wooden instrument. He described his results in patients to his colleagues. He contrived words to represent the sounds that he heard: rales, crepitations, murmurs and many others. When a patient died, the autopsy was correlated with the clinical findings. He established the anatomical significance of both normal and abnormal chest sounds that are still in use today.

The stethoscope was the first diagnostic instrument to achieve rapid international popularity. Within a short time, numerous foreign students transported the tool across Europe and to North America. Anatomy had suddenly been made to fit into the clinical practice of medicine.

A second edition of Laennec's book with all his observations was released in May 1826 in it he responded to criticisms of his new invention. But three months later, Laennec was dead of tuberculosis at age 45. During his terminal illness physician friends examined the inventor with his own stethoscope, but they concealed their findings of advanced disease in order to boost his spirits in his final days.

Laennec is remembered because he had employed his ears to see inside the body—visualization of internal structures and inner workings was his goal. The process of using a stethoscope continues even with all the advanced technologies now available.

Understanding the Digestive Process—William Beaumont

Paralleling the new role of the stethoscope was a new creative breed of scientists looking at the basic functions of the body—the study of physiology. One area of mystery was digestive activity. No one understood how the stomach processed food. Was it ground up by the stomach, heated so that it melted into a liquid form, or was it changed chemically?

Military Surgeon William Beaumont (1785-1853) was presented with a unique opportunity to study digestion and answer questions about the role of the stomach in the process. He was the first American to attract international fame for his physiological investigation of digestion in a very unusual case.

In 1822 while at Fort Mackinac, Michigan fate intervened to change the life of Beaumont and a young patient, French-Canadian fur trader Alexis St. Martin. While at the fort for supplies, a musket accidentally discharged hitting St. Martin in the abdomen. Beaumont rushed to the scene and did all he could to repair the wound. Amazingly St. Martin survived, but he was left with a permanent external fistula (a hole) to the stomach.

It occurred to the inquisitive Beaumont that he might be able to learn about digestion by studying the French-Canadian. With his patient's permission the doctor began a series of experiments by tying a string around pieces of meat and other foods, inserting them through the fistula and retrieving them after varied periods of time. He carefully documented how the gastric juices worked on the different types of food and to what extent the amount of time within stomach played in the process.

So eager was Beaumont to maintain the investigations over the next decade that he often had St. Martin live with him for up to two years at a time. In 1833 he wrote about his observations and the knowledge he acquired. It was a landmark paper that changed the way physicians thought about the stomach's role in digestion. Shortly after the appearance of the publication St. Martin had enough of Beaumont and returned to his hometown in Quebec Province, Canada.

Sadly, Beaumont died in 1853 before his patient's passing. St. Martin died at age 86 in 1880, 58 years after the gunshot accident. Because St. Martin's family did not want any publicity about his role in Beaumont's experiments, he was buried in an unmarked grave next to a Roman Catholic Church. In 1962, the Canadian Physiological Society placed a bronze plaque on the church's wall to commemorate St. Martin's contribution to science and medical history.

As often is the case with new discoveries, the role of serendipity frequently plays a part. Beaumont recognized an opportunity to learn and observe a heretofore unknown activity in his unexpected encounter with St. Martin. An aphorism of another famous 19th century scientist said it best "in the field of observation, chance favors only the mind that is prepared." Beaumont's brain was ready. And so was the mind of the man who said that quote, Louis Pasteur.

The Germ Theory of Disease:
Louis Pasteur and Robert Koch

For many the germ theory of disease was the single most important discovery in medicine in all times and all places. While this may seem like an exaggeration, remember that infectious diseases were historically the most significant cause of morbidly and mortality since the time of the caveman. Recognizing that microbes could cause illness was a centennial event that debunked past conjectures like miasmas and the four humors as the basis for maladies. The principle architects of the germ theory and microbiological revolution were Louis Pasteur in France and Robert Koch in Germany.

Pasteur was the superstar of the 19th century generation of medical researchers. He was the son of a tanner and not a physician but rather trained as a chemist. This may explain why he approached medical problems in a basic, scientific fashion, rather than reaching out immediately for practical answers.

His breakthrough research began with his studies of spoilage or souring problems in his country's wine industry. First, he demonstrated conclusively that fermentation was the work of various microorganisms and not a purely chemical event as was the prevailing theory. Second, he proved that the souring of wine and beer was caused by contamination with bacterial microbes.

In 1864 he devised a remedy to stop the wine from going bad. He heated the wine briefly to between 122-140 degrees to kill off disease-causing bacteria. The increased temperature did not alter the beverage's aging process, taste or appearance. In the 1880s this process became known as "pasteurization" in his honor. As a public health measure, it helped save many lives over the years by preventing the spread of disease through contaminated milk.

Pasteur continued to work exclusively with microorganisms. He developed methods to cultivate them, learned their nutritional requirements, and gradually implicated them in animal and human diseases. His discovery that the injection of the dried microorganism culture of chicken cholera could actually protect the animal from disease led him to his triumphal work on vaccines. He adopted the name—vaccines—in honor of Jenner's smallpox vaccine.

All this led to Pasteur's analysis of anthrax. In the late 19th century it caused a deadly and highly communicable disease in farm animals. Pasteur found that by heating the anthrax bacillus to reduce its virulence, he could produce a vaccine.

Its efficacy was proven in a public demonstration in a farmyard outside of Paris in 1881. Pasteur first injected 24 sheep with his new vaccine. Then a month later witnessed by newspaper reporters and other invited observers, he injected the same vaccinated sheep as well as an additional control group of unvaccinated animals with the virulent anthrax ba cilli. The results were dramatic. The vaccinated sheep stayed healthy and the unprotected ones died even as the reporters arrived on the scene. It was a well-orchestrated public relations success for Pasteur and a victory for the science of bacteriology establishing it to be a new indispensable branch of medicine

Pasteur's encore was even more audacious. The disease he sought to treat—not simply to protect against—with a vaccine was not caused by known bacteria, but rather by an agent invisible at the time, a virus. The affliction was rabies. It was not a common human infection, but when present it was ghastly, painful, and universally fatal.

Pasteur was greatly puzzled by the disorder because he could not isolate or culture the agent responsible for the illness. This is not surprising since a virus is now known to be the cause and could not be visualized with a light microscope. It can only be seen with an electron microscope. The rabies' virus was first visualized in 1931. Even with this handicap of not identifying the etiology, Pasteur learned how to attenuate the mysterious causative agent and was able to produce a vaccine to treat the dreaded disorder.

On July 6, 1885 a 9 year old boy was bitten multiple times by a dog thought to be rabid. His family physician sent him to Pasteur who took a great risk. For the first time Pasteur gave a human, the young boy, a series of 14 painful injections of his vaccine. The child remained well and did not develop the symptoms of rabies. After several other trials, his rabies procedure became the standard of care. Over the next 15 months, the preparation was successfully given to 2,000 people and many more during the next decade. His rabies antidote was likely the apogee of 19th century laboratory medicine.

A grateful French nation built Pasteur a special research institute in 1889. He was buried there on his death six years later. He was remembered for his restless, creative genius, and for his great human kindness. In the end, although he was never a physician, his work helped save countless humans and helped initiate the role of experimental bacteriology and microbiology into clinical medicine.

Pasteur was a wizard, both within the lab and beyond, but bacteriology's consolidation into a scientific discipline was due mainly to German Robert Koch (1842-1910). His painstaking microscopic work established the germ concept of disease and systematically developed its potential. In contrast to Pasteur whose road to microbiology began with chemistry, the younger Koch came to bacteriology as a physician. His scientific investigations were primarily motivated by medical questions. He had the advantage of having grown up with the benefit and knowledge of the decades of Pasteur's experimental and microscopic research.

Koch's approach was different. He was a bacteriologist, not a microbiologist like Pasteur. Koch was focused on the role of bacteria in the cause of human disease. He identified many in his years of work. In the process, Koch frequently attacked

Pasteur's findings openly and directly much to the chagrin of many other scientists. He even tried to discredit Pasteur by saying he was just a chemist, not a medical doctor. Many speculate the cause of this rivalry initiated by Koch was not all scientific in origin but rather based on the hatred and prejudice that the German people had at that time towards anything that was French. Throughout his productive career, Koch pressured himself to match the achievements of his great French rival.

Koch's achievements covered a wide clinical and scientific spectrum. Two different areas highlight and serve as examples of his creativity in one instance and on another occasion a wayward motivation that destroyed his scientific credibility. They are his postulates and his investigation of tuberculosis.

In his search for identifying the etiology of various diseases, Koch was meticulous in deciding whether an isolated bacterium was the cause of a particular infection or simply background noise. In 1883, he developed a formulation—a set of rules—to help in the differentiation. It came to be known as "Koch's Postulates." A modified form is still used today. They are:

- The suspected germ must be consistently associated with the disease.
- It must be isolated from the sick person and cultured in the laboratory.
- Experimental inoculation with the organism must cause the symptoms of the disease to appear.

- In 1905 a fourth rule was added that said that an organism must be isolated from the experimental infection.

These provisions for the etiological identification of various maladies could mostly be filled except notably in the case of viruses. In these instances, viruses were accepted as causes without meeting all the rules' stipulations. Other scientists adopted these provisos to figure out cures for different diseases. The postulates were formulated for infectious diseases. The general approach and guidelines were also extended to investigations of other associated medical problem, like the health hazards posed by asbestos and other chemicals.

In 1881, Koch began his work on tuberculosis. It took him on a very different path than with his postulates. It ended with a less than honorable conclusion. He committed all his energy to the task of identifying the cause and cure for this ubiquitous malady. Unlike most of his other investigations, this time he functioned in strict secrecy.

To comprehend why he was so dedicated to this undertaking requires an appreciation of the ways in which this disease permeated the fabric of life throughout Europe in the 19th century. Tuberculosis was, in terms of the number of victims claimed, more devastating than most dreaded epidemic diseases like smallpox and cholera. Called the "captain of the men of death" in the 17th century, by the mid 1800s tuberculosis was the cause of about one in seven deaths. Its impact on society was greatly amplified because it was more likely to infected those in their most productive adult years.

Prior to Koch's research, 19th century author Charles Dickens in 1870 expressed the despair of many about tuberculosis by characterizing it as "the disease that medicine never cured and wealth never warded away."

In just under a year Koch made a dramatic breakthrough. He cultured a specific microbe, the tubercle bacillus. He provided solid evidence from animal experiments that it was the cause of the disease. Subsequently, perhaps in a need to eclipse Pasteur with one great therapeutic coup, Koch went back to the laboratory once again with increased intensity and continued secrecy to find a countermeasure.

In 1890 Koch revealed that he had discovered a substance that arrested the growth of the tubercle bacillus in the test tube and in living bodies. He called it tuberculin. He proclaimed that it was a remedy. This led the world to believe that he had a cure for tuberculosis. At the time of the announcement he avoided disclosing the exact nature and composition of his clandestine tuberculin substance.

Within a year, thousands had received tuberculin. It seemed to help some in the first stages of the disease, but experience quickly showed that it was useless or even dangerous to those already with pulmonary tuberculosis. The fiasco brought a violent backlash. There were denunciations of Koch and his sub-rosa antidote. In January 1891 he was forced to reveal that his cure was nothing more than a diluted glycerin extract of the tubercle bacillus itself. He was accused of finally divulging the makeup of his secret countermeasure only after it had become an obvious ineffective remedy and financially worthless. Under a cloud of suspicion Koch disappeared to Egypt with his young bride. He left his underlings to cope with the debacle.

One positive result of this disaster was that a diagnostic test using the tuberculin material was designed. It became the standard means of demonstrating if someone has been exposed to the tuberculosis causing microbe. With no cure on the horizon, victims of the disease well into the 20th century were isolated, shunned and confined in sanatoriums in a manner reminiscent of the medieval leper. In these institutions those afflicted were treated with rest and improved nutrition.

Before an effective antibiotic was found, surgical treatment of tuberculosis was common and could be lifesaving. By draining pleural fluid from around the lung of infected patients, life was prolonged. In addition, a "plombage" technique that collapsed the infected lung to allow it to rest was used. But surgery lost its efficacy with the introduction of drug therapy.

In 1943 the antibiotic streptomycin was discovered and was established to be a potent measure against tuberculosis. A more beneficial drug, isoniazid, emerged in the 1950s. While usually curable, today tuberculosis still affects 8-10 million people every year in underdeveloped nations. The tubercle

Jonathan L. Stolz, M.D.

bacillus has mutated creating drug resistant strains making treatment a continuing challenge and a global emergency once again.

As for Doctor Koch, he overcame his tuberculin failure. But he is not well known today. In fact, his assistant, Richard Petri developed a special plate for use with agar cultures. Thanks to the universal use of the 'petri dish,' Petri is generally more familiar to contemporary biology students than his boss Robert Koch.

Antisepsis: Ignaz Semmelweis and Joseph Lister

Pasteur and Koch's work inspired Scottish surgeon and experimental scientist Joseph Lister (1827-1912). In 1867 he introduced his antiseptic system into the practice of surgery. He was not the first to recognize the high mortality of wound infections in hospitals. Almost a generation before Lister, an obscure Viennese obstetrician, Ignaz P. Semmelweis (1818-1865) discovered a method to prevent childbirth fever. This affliction had similar characteristics to post-operative wound infections.

Semmelweis' life encompassed elements of both heroism and tragedy more appropriate to treatment by a novelist than a medical historian. While working in two separate obstetrical clinics in Vienna in 1847, he was struck by the marked differences in mortality between the facilities. One was staffed solely by midwives, and the other by doctors and medical students. His analysis showed that the site with the greatest number of deaths was the result of unwashed contaminated hands of the doctors and medical students who proceeded directly from performing an autopsy to deliver a baby.

He introduced hand washing with a chlorine solution in the birthing room. The result was a dramatic reduction of puerperal mortality. Unfortunately, his colleagues ignored Semmelweis's findings and advice. He was even dismissed from his position because of his discovery and hand washing recommendation.

For the remainder of his life Semmelweis preached with zeal his new principle, but he was not heard. His mind became increasingly confused in his frantic fight against the intransient conservatism of his colleagues. In 1865, two years before Lister's antisepsis was announced, Semmelweis died at age forty-seven from a toxic infection in an insane asylum where he had been hospitalized because of his bizarre behavior. In the 1880s when Lister eventually became aware of Semmelweis's work, he generously gave him credit for introducing the principle of asepsis into surgery.

Even with the debut of anesthesia in the early 19th century, modern surgery would not have evolved without Joseph Lister's introduction of his antiseptic system. Lister's appreciation of Pasteur's germ theory and his obsession with preventing infection for both the surgical operation and post op care was the key to his success. His technique was the use of carbolic acid and other antiseptics. They were splashed into wounds or sprayed into the air to kill germs that it was presumed were lurking in the atmosphere.

It took time to overcome the skeptics of this new method. By the end of the century most surgeons embraced Lister's antiseptic measures that killed the germs in and around a wound by means of germicidal agents. Many soon then realized that surgical wounds should be clean from the outset. Preventative asepsis was introduced to avoid wound contamination by surgeons and instruments in the operating field. The autoclave and rubber gloves soon became a part of the full preoperative aseptic ritual. Today, additional precautions within the operating room theater have made postoperative infections rare. When Lister retired in 1892, he had won deserved recognition. He was the first British surgeon to be elevated to the peerage.

Surgery in the United States

The practice of surgery from the last half of the 19th century to the first several decades of the 20th century underwent a transformation. The work of Joseph Lister in England was a pivotal development in this change. American surgeons accepted some aspects of Listerism, but it was not until the 1890s before all of Lister's principles were fully embraced.

As with many other medical developments surgery required hospitals and a large population to become a specialty in its own right. This transition in the late 19th and early 20th centuries in the United States was initially problematic. America was still largely rural, and most operations was performed by physicians in private homes. Although a few large hospitals and urban centers provided some opportunity for surgery to develop as a specialty, the United States had nothing comparable to the great European medical centers.

By the end of the 19th century with acceptance of aseptic techniques, a new generation of American surgeons appeared on the scene. These physicians were better educated than their predecessors. Nearly all had studied in major European medical centers. This was the age in which pioneers in the specialty developed historic new techniques to deal with traditional problems.

William Halstead at Johns Hopkins was an early innovator. He introduced rubber gloves into the operating room. He performed the first radical mastectomy for breast cancer in 1882 and created novel methods to decrease damage to the blood supplies to tissues during surgery. Halstead initiated the first formal surgical residency program at Johns Hopkins University. Some credited him with being the father of modern surgery. Unfortunately, he became addicted to cocaine that ended his brilliant medical career.

Outside the achievements of academic surgery, the unassuming appendix became a focal point. It encapsulated the progress that was made in clinical practice and surgical technique. Until the 1880s abdominal surgery of any kind had been one of medicine's taboos. The operative death rate was high. In 1894, New York surgeon Charles McBurney greatly facilitated the diagnosis of appendicitis. He identified the classic area of tenderness localized in the right lower quadrant that indicated the presence of an inflamed appendix. It is still known today as McBurney's point. He subsequently developed a simplified operative technique to remove the infected painful organ.

The increase in technical skills that surgeons obtained by performing appendectomies resulted in a marked decrease in the death rate from abdominal surgery. Patients' willingness to undergo appendectomies increased as physicians improved their diagnostic skills and lowered their operative mortality rates. By 1930 ten percent of patients admitted to American hospitals had appendectomies.

Appendicitis became somewhat of a fad and a fashionable disease in popular culture. One physician wrote, "People became so fidgety about their appendix that it only required the consumption of a few green apples to send them pell-mell to the hospital." Some authors believe this surgical treatment helped in a small way to redefine medicine. Physicians became proactive about treatment with public awareness campaigns that emphasized early detection like "Don't gamble with appendicitis, don't use a laxative, call a doctor." As for the appendix itself, the cause of all the excitement, it is a vestigial organ. It has no real medical purpose except perhaps some immunological benefit for the patient .

The early general surgeons who benefited from performing many types of operations warned against the rise in specialization, but to no avail. Ophthalmology had long been a separate field. The introduction of the cystoscope with an electric light for bladder surgery paved the way for urology. Harvey Cushing the outstanding leader in cranial surgery during the early years of the 20th century helped establish the basis for making neurosurgery a recognized specialty. Other areas soon developed.

CHAPTER SEVENTEEN
New Pharmaceuticals
and Diagnostic Technology

At the end of the 19th century, pharmaceuticals were transformed by the breakthroughs of Pasteur, Koch, and others. This was especially welcome because drug development had lagged behind other branches of medicine—hence the doctrine of 'therapeutic nihilism' became well known.

It was the knowledge of germs that was the catalyst for new studies in immunity and alternative ways that infections could be prevented or overcome. This led to the development of additional vaccines and antitoxins and drugs derived from chemicals. They were by modern standards, crude and made from impure materials. These two forms of treatment were known as serum therapy and chemotherapy.

Serum Therapy

Serum therapy was based on the notion that natural bacterial antitoxins in the blood could neutralize bacterial toxin present in the body. The process involved having an antitoxin produced by experimental animals that then could be utilized to immunize or even cure infected individuals.

This toxin-antitoxin concept produced a new class of therapy. An example of an illness successfully treated by this method was the once common childhood disease of diphtheria. It is an acute infection caused by the toxin-producing *Corynebacterium diphtheria* that was discovered in 1883. Inhaling the bacteria generated the disease. Within a week after being infected, the victim experiences generalized illness and a characteristic "false membrane" in the throat that could cause death by suffocation.

Case fatality rates for diphtheria rarely exceed 10%, but historic epidemics took such a heavy toll that the disease was still feared at the beginning of the 20th century. A diphtheria toxin was introduced in the last decade of the 19th century. It was only one of several bacterial toxins to have found a valuable place in therapeutic medicine.

In the 1990s diphtheria toxin was combined with a protein that binds to certain white blood cells. It served as a prototype for a genetically engineered drugs and a new approach to treat rheumatoid arthritis and some other diseases.

Early Chemotherapy

By the turn of the 20th century there was a growing symbiosis between drug researchers and manufactures in Europe. The booming chemical industry realized that significant profits could be made in producing and then selling pharmaceuticals. These firms partnered with academic pharmacology departments particularly in Germany. Here one of the leading researchers was Paul Ehrlich (1854-1915).

His idea, that he termed chemotherapy, was that simple chemical substances might act as powerfully agents against microbes without harming the patient. The challenge was to find synthetic chemical substances lethal to a particular organism but non-toxic to its host. He termed these drugs "magic bullets."

Ehrlich tried his scientific concept by addressing the cure for that pervasive ever shocking disease syphilis. In 1905 the protozoan parasite that caused the malady was isolated. Then in 1906 a diagnostic screening test was developed and named after its founder August von Wasserman.

Ehrlich was a tireless worker. He believed the keys to success in research were "money, patience, and luck." Along with his Japanese colleague Sahachiro Hata (1873-1938) they systematically tested hundreds of different chemical varieties of arsenic and other amalgamations. Finally, after years of investigation, the 606th compound examined was found capable of killing the spirochete of syphilis without being too toxic to the patient. He named it Salvarsan. It was the first chemotherapeutic agent specifically aimed at the microbe that causes syphilis.

By 1910, when the drug was formally marketed, Ehrlich was a sick man. For the remaining five years of his life he spent his time modifying the medication to a better efficacy/toxicity ratio as well as defending it against anti-Salvarsan crusaders. Those opposed to its usage claimed that the treatment had many side effects and should not be marketed to the public. These antagonists ignored the fact that over one million people had been successfully helped.

Ehrlich died of a heart attack in August 1915 knowing that his drug Salvarsan was the first "magic bullet." Many more would follow in the future. Salvarsan remained the standard remedy for syphilis until it was replaced by penicillin after World War II.

Aspirin

Late in the 19th century, one of the most significant discoveries was the simple and effective aspirin. In ancient civilizations people knew there were substances in nature that could be used to ease aches and pains. One of those dating back to the time of the Egyptians was the juice from the bark of the willow tree. Many years later scientists in the mid 19th century studied the chemical makeup of this well-known cure all. Its composition became known as *salix*, which is Latin for willow.

The next challenge was to make this chemical compound into something that could be proffered as a medication. The initial results from an Italian chemist produced salicylic acid that worked to reduce swelling and pain. Unfortunately because of a severe side effect that produced significant bleeding in the gastrointestinal tract, it was abandoned as a treatment.

In 1897 Felix Hoffman, a German chemist employed at Bayer & Company, reduced the acidity of salicylic acid by converting it to acetylsalicylic acid or ASA. After some initial reluctance on Bayer's part to introduce Hoffman's discovery it was finally marketed in 1899. It was called aspirin. "A" was for acetyl chloride, and "spir" was the spirea ulmaria plant from which salicylic acid was derived, and "in" was added as a common ending for medications.

Bayer was the ideal company to be the first major manufacture. From their experience as a dye company they were accustomed to a business model that involved mass production and marketing on a scale far beyond what most other drug companies had done in the past. Bayer was extremely successful with its new drug, but its exclusivity ended after World War I. As a German company part of the reparations in the Treaty of Versailles gave Bayer's trademark on aspirin as well as on heroin to France, England and Russia.

Aspirin is now used for more than the reduction of pain and fevers. It is a truly versatile drug. Its protective effect against heart disease and reducing the incidence of some forms of cancer is now well known. It was not until 1982 that scientists grasped how aspirin actually worked to reduce pain. It is the drug's effect on prostaglandins (hormone like chemicals in the body) that makes it potent. Aspirin does not actually correct or cure a headache or arthritis, but what it does is to decrease the number of pain signals traveling through the nerves to the brain.

It is a true miracle drug that was commercially introduced over a century ago. With all aspirin's known potential side effects now understood, it would be unlikely that the present day United States Federal Drug Administration would approve its use for patients if it were a new application.

Early Diagnostic Technology

While there were changes in drug therapy in progress, other clinical scientists looked beyond the 1819 introduction of the stethoscope for new diagnostic tools. Several important diagnostic aids were introduced. Most were technologically sophisticated, labor intense, and suitable for use only in the hospital setting or by the physician specialist. This limited their penetration into the general medical practices.

The ophthalmoscope was one of the earliest of these instruments. It typifies many of the features of the diagnostic technology of the period. The modern instrument was introduced into medicine in 1851. It had a direct impact on clinical medicine. It helped ophthalmologists evaluate the cause of defective vision by direct visualization of the retinae, optic nerve head and the arteries that supply the eye. From its debut it value and use remained within the provenance of consulting physicians.

Other 19th century tools allowed doctors to detect abnormalities inside the body. The otoscope and laryngoscope made visualization of the inner ear and the throat, respectively much easier. These instruments were simpler devices than the ophthalmoscope, easier to use even if the interpretation of what the doctor saw was always a learned skill.

Blood pressure values could be measured and recorded by the invention of the sphygmomanometer in 1881. Electrocardiograms in 1903 advanced the understanding and diagnosis of cardiac abnormalities like angina and myocardial infarction.

X-Ray

All these technologies permitted the physician to peer beyond the patient's history and symptoms and look into the body to identify the material basis for the illness. The most dramatic change in diagnostic capability was the discovery on December 28, 1895 of X-Rays by Wilhelm Conrad Rontgen (1845-1923). Rontgen was a German physicist who stumbled on the properties of cathode rays' remarkable ability to penetrate many substances, including human flesh. He accidentally exposed his wife's hand on a photographic plate with the machine he was working on at the time. The result was an image of the bones of her hand. This was a momentous medical event that became rapidly known as the news spread throughout the world. When the New York Times broke the news in January 1896 the article concluded that it would transform "modern surgery by enabling the surgeon to detect the presence of foreign bodies."

Later that year the first diagnostic radiographs were generated in the United Sates at Columbia University and the University of Pennsylvania. The development of the potential medical applications for x-ray was essentially an international accomplishment. It was one that American physicians played a significant role. During this time fluorescent screens were invented that allowed the radiologist to see a patient's image in real time for use in contrast exams like barium studies. This early technique was a hazard because the doctor had to look directly into the x-ray beam coming from the fluoroscope. Many cataracts developed prematurely in these early radiologists.

Before x-ray's medical usage became apparent, the public was infatuated with what these unknown rays could see. They were promoted in the entertainment field and had influence throughout society. There was even a concern that the x-ray would be a danger to the conservative Victorian values because it exposed women's bodies. A poem written in 1896 reflects this thought:

> *The Roentgen Rays, the Roentgen Rays,*
> *What is this craze?*
> *The town's ablaze*
> *With the new phase*

Of X-ray's ways.
I'm full of daze,
Shock and amaze;
For nowadays
I hear they'll gaze
Thro' cloak and gown—and even stays,
These naughty, naughty Roentgen Rays

After this initial fascination, within a decade, some physicians began specializing in x-ray procedures. For the next fifty years radiology grew at a rapid pace. During these decades techniques were generally confined to examinations that involved creating an image by focusing the x-ray beam through the body part of interest onto a film inside a special cassette. In the early years it could take up to eleven minutes of x-ray exposure to generate a single film to interpret. Today it takes just milliseconds to produce similar routine radiographs .

It was the digital age that allowed radiology to explode into new technological advancements like ultrasound, Magnetic Resonance Imaging and Computed Tomography scans. These revolutionary imaging modalities would not have been possible without the invention of the computer. Many patients have benefited from this progress by reducing the need and risk of exploratory surgery that was common up until the mid 1970s.

CHAPTER EIGHTEEN
Medical Education Transformation

In the mid 19th century a young man seeking a medical education might have found himself—women were rare in the field at that time—in one of four settings depending on his country of origin, financial resources, and career ambitions.

The apprenticeship was the oldest and most basic. Here a mentor was found and practical knowledge was learned in an informal non-academic environment. With the new demands of medical knowledge, it started to decline in influence and slowly disappeared by the end of the 19th century.

Proprietary schools were run by an individual or a group offering some basic instruction in medical subjects. While present in England and Europe, in laissez-faire America these schools were nearly the entire the medical scene—not just as an adjunct to supplement regular training but also for the total medical education experience. Some were run by conscientious proprietors, but the majority were concerned primarily with profit. Their profusion in the United States contributed to a surplus of doctors in the mid 19th century and to the low reputation that American medicine had in Europe. These schools were the chief animus of Abraham Flexner's (1866-1959) early 20th century evaluation of American medical education.

The third type of medical school was that associated with a hospital. This was most common in England and France. Here the student's curriculum was generally of a practical nature. Graduates were prepared for the realities of the marketplace. For many aspiring future leaders within the profession, this was not enough.

The gold standard of medical education was the university medical school. Only there could the student receive instruction from those who were actively engaged in research. European schools dominated this type of education in the 19th century.

United States Medical Schools & Flexner Report

In North America, the formally educated physicians from the colonial period to near the end of the 19th century received all or a part of their training in Europe. The first American medical school opened in Philadelphia at the predecessor of the University of Pennsylvania in 1765. Others followed shortly thereafter at Harvard, Columbia and other large university centers. These institutions only trained a minority of American doctors. By the end of the 19th century the majority of United States medical schools were proprietary establishments with no research facilities. The need to recruit students' tuitions took precedence over any entrance require-ments or scientific advancements.

The early 19th century view of medical students as a crude and unruly group carried down to the end of the century. One physician teaching at New York University described the lec-ture hall as "an ill-constructed dirty room drenched with tobacco and perfumed with vile odors." At the end of the class hour he observed that there was the "most excruciating noise that splits your ears as the students rushed forth like mad buffaloes." The growing awareness of the inadequacy of medical training led to the establishment of many state licens-ing boards. Since these agencies tended to accept almost any medical degree, the net effect was to stimulate the rise of inferior schools. The number of medical colleges increased from 90 in 1880 to 151 in 1900.

With laboratory science in the later decades of the 19th century becoming crucial to medicine, it took a new cohort of American physicians to reform medical education. Almost all of these doctors had received some of their training in the progressive European medical schools where research and scientific laboratories were a part of the educational process. When American students instructed at these universities re-turned home, many joined the faculties of Harvard, University of Pennsylvania, Colombia and others. They became the catalysts pushing for the adoption of European educational ideas into the curriculums of their medical schools.

One of the first to initiate reform was Harvard in 1869. Charles Eliot was President of the university and one of the new generation of men trained in science. He recognized the inadequacy of his medical school. He forced a change that lengthened the school year and implemented a longer three-year curriculum. Both written and oral exams were required in order to graduate. These alterations resulted in a decrease in the number of students at Harvard by 43% between 1870 and 1872. But the improved quality of the educational process was soon recognized. The University of Pennsylvania, Syracuse, Yale and Michigan followed in also establishing a three-year curriculum.

While older university medical schools were striving to raise their standards, it was the founding of Johns Hopkins Hospital in 1889 and its medical school in 1893 that profoundly transformed all American medical education. At Hopkins, there was a sophisticated level of scientific research carried out at the medical school that was fully organized and integrated into the academic workings of the larger university. This new standard for graduate education stimulated radical change in all medical instruction for future physicians. The man responsible for this reform at Johns Hopkins was William H. Welch.

Like many in his generation Welch, after graduating from Yale and Columbia medical school, traveled to Germany to complete his training. He was a visionary, insightful, and imaginative. Drawing on his exposure to German universities, he realized that important scientific breakthroughs would occur only if American doctors related laboratory research to their clinical practice. This was the foundation of Welch's success.

The opportunity for Welch to put his dream into a real world model came from the trustees of Johns Hopkins University. They invited him to move to Baltimore to implement his ideas. He recruited staff for their new hospital and developed a curriculum for the medical school. In a short period of time William Halstead came to head Surgery and William Osler as Professor of Medicine.

Welch and his colleagues initiated a quantum leap in medical teaching. It was the first time a bachelor's degree was required for admission to medical school. A Johns Hopkins medical education meant learning by doing. The intensive laboratory work of the first two years mitigated the tedium of basic scientific instruction. The around the clock bedside care of the last two years ensured that the academic medical knowledge developed into professional and practical competency. No other medical school at that time had this level of medical education.

Although not the initial school to build a teaching hospital, Johns Hopkins was the first to make the healthcare facility an integral part of a research oriented institution. Within a few years the school's graduates and younger faculty members extended far beyond Baltimore. They became a major factor in shaping the character of medical education and research in the 20th century in the United States.

While the world at large celebrated Welch's medical achievements, he was known in Baltimore as much for his personality as his professional prowess. His enthusiastic and paternalistic nature led friends and students alike to nickname him "Popsy." He was a bachelor with a hearty laugh. He was an engaging conversationalist that enjoyed the pleasures of eating out and delighted in being a memorable and somewhat of an oddball within the community. Unlike many characters in American medical history that outlived their contributions and died in anonymity, Welch remained influential all his life. On his 80th birthday in 1930 he appeared on the cover of Time magazine.

Sharp contrasts characterized medicine by 1900. The changes at Harvard, John Hopkins and elsewhere were counter-pointed by the continuing growth of the proprietary/commercial medical colleges. At these schools and weaker universities, the ranks of the profession were being recruited from the lower middle classes much to the dismay of professional leaders who thought such riff-raff jeopardized efforts to raise the doctor's status in society. Medicine would never be a respected profession for these regular university trained professionals until the perception of the course and common elements within the physician community at large were reduced.

Into the fray stepped the American Medical Association (AMA). It was founded it 1847 by orthodox allopathic physicians that were opposed to non-traditional doctors like homeopaths and others they considered practicing quack medicine. The AMA members collectively sought to raise the standards of care in order to benefit the traditional doctors. By 1901 it was apparent that the organization was only partially successful in its effort to influence national medical policy. So it reorganized and emerged as a much stronger advocate for the medical profession.

During most of the 20th century the AMA became the central group that spoke for practicing physicians. The organization's leaders articulated authoritatively on a variety of health issues and vehemently opposed any programs that interfered in the relationship between the doctor and patient. At its peak AMA membership in the 1950s included nearly 70% of practicing physicians. Today, they are still the largest physician advocate group, but because of the growth of specialty societies they now represent just over a third of all doctors in the United States. Nonetheless, the AMA's influence remains as a formidable lobbyist on behalf of practicing physicians.

In the beginning of the 20th century one of the AMA's initial projects was the formation of the Council of Medical Education. Its purpose was to formulate new standards for education and medical training. In 1906 an AMA delegation inspected the 160 medical schools then functioning. It fully approved only 82, but the AMA never publicly published the entire results of their survey fearing that the ill will it would

create within their membership. The association's professional ethics forbid physicians from taking a public stance
against each other.

To make their findings more widely known, the AMA decided to invite an outside group, the Carnegie Foundation, to
conduct a similar investigation. The foundation hired a young
educator, Abraham Flexner, to produce the report. Flexner
had earned his bachelor's degree at Johns Hopkins University
where he became a great admirer of William Welch. Flexner
began his task by initially examining in detail the educational
curriculum and teaching structure of the Johns Hopkins
Medical School. He decided to use its template as the
benchmark for measuring all medical schools in the United
States.

His published report in 1910 gave a unique and important
perspective on medical education. It was a pivotal document
in which all American medical schools were evaluated.
Flexner made a complete and detailed analysis of every school
then in existence after personally visiting everyone. Though a
layman he was more severe in his conclusions about the
scholastic capacity of each facility than the AMA had been in
its survey. In firm and decisive prose, he passed judgment on
each school, dividing them into three distinct groups.

Those at the top like Harvard, Michigan, Penn offered the
best level of medical education that the country had at the
time. At the bottom of the group were the majority of the
proprietary schools. They would require an enormous

amount of work to reach the standards of excellence predicated by Flexner. Finally, in the middle of the pack were institutions that needed some support to remain viable. The overall result was to widen the gap between the better and poorer schools. This caused most of the proprietary schools to gradually fall by the wayside. By 1930, twenty years after the report the number of medical schools in America dropped from the 148 surveyed in 1910 to 66.

As a spinoff from his publication, Flexner generated widespread interest in medical education that served to inform the public on the importance of scientific medicine. He also helped convince private philanthropists and the government to fund research oriented medical schools. The Flexner report truly advanced American medicine forever.

Between 1920 and 1960, few new medical schools appeared in either North America or Europe. But as the graduate-per-capita ratio slowly declined in the mid 20th century, surveys demonstrated the need for more doctors. A boom in creating medical schools took place. Today, there are 126 allopathic medical school and 26 osteopathic schools in the United States.

Women in Medicine

Women everywhere in the world have always played a broad role in healthcare from local healers, family caregivers to midwives and nurses. Throughout history the male dominated

medical profession has ignored the gender imbalance and discrimination against females. What took place in the United
States is illustrative of the plight women have had in
medicine. .

Prior to 1849, no American woman had graduated from
medical school. This changed that year when Elizabeth
Blackwell received her medical degree. By the Civil War at
least three medical schools exclusively for women were grant-
ing degrees and some of the proprietary schools in the post-
Civil War era became coeducational. The opening of Johns
Hopkins Medical School helped by admitting a woman to
their first class in 1892. With the woman's rights movement
in the late 19th and early 20th century a highly articulate minority of male physicians supported the women's cause.
Unfortunately, the voices raised in opposition to female doctors were more numerous and strident. This led to a decline
in women interested in becoming doctors.

In the first 60 years of the 20th century there were a variety of
factors that limited the entrance of women into medicine.
Social and economic issues included parents' reluctance to
spend money on their daughters' education and the notion
that graduate schooling for women reduced their chances of a
successful marriage. Likely the most telling reason was the
deliberate effort by male medical staff faculty and administra-
tors to discourage women students. For most of the 20th
century, medicine was considered and accepted as a masculine domain.

This situation began to reshape in the late 1960s when social activists sued public institutions to deal with the systemic discrimination against women. Between 1972 and 1980 American medical schools almost doubled the number of females admitted from 15% to 28%. By 2005 the number of women entering medical school equaled the number of men—that was a 400% increase in 35 years. Today, women currently make up over one-third of the total physicians in the United States. Although females are now better represented in the profession as a whole, there remains an imbalance within the specialties. They are particularly underrepresented in the surgical specialties and procedure oriented fields.

CHAPTER NINETEEN
Medical Progress
World War I Through World War II

Medicine Advances in World War I

Medicine advanced during the second half of the 19th century. The level of medical knowledge in 1850 was vastly different than that practiced at the outbreak of hostilities in 1914. In those 64 years the basic sciences all came together as a unifying force. Because of this medicine played a more important role on the blood soaked battlefields of World War I than in any other previous conflict.

The United States and the allies' involvement in the war required that governments utilize both curative and preventative medicine to the fullest extent. The variety of health professionals mustered included the usual doctors, nurses, and pharmacists. Unlike in previous wars, sanitary engineers, lab techs and physician specialists of every stripe were also recruited.

Testimony of their collective value came after the conflict when the statistical record revealed that for the first time in history, there had been fewer deaths from disease than from battle wounds. This remarkable accomplishment—before the era of antibiotics—was achieved by innovations in a number of areas. These included more thorough examination of recruits, education and prophylaxis against venereal disease, and improved enforcement of sanitation and hygiene.

The wounded soldiers were taken to hospitals from the battlefields in a better system of triage and evacuation. The treatment of the enormous number of causalities employed science based therapies. In the military hospitals surgeons undertook a far greater variety of complex operations than had their predecessors in previous wars. Doctors used newly developed antiseptic solutions to irrigate injuries and refined a fundamental surgical principal—the removal of all devitalized tissue prior to suturing.

Blood transfusions that had been used sporadically prior to the fray became a reality. They had been first tried in America in 1832. The failure rate was high and awaited advances in the basic sciences. This came at the turn of the century when Austrian physician Karl Landsteiner provided the answer for successful transfusions. He noted that while all human blood may visually appear the same, it actually had variable characteristics that he referred to as blood types. It is now well known as A, B, and O groups. Other scientists distinguished a fourth main blood type, AB. A United States Army Medical officer demonstrated in 1917 that blood could be donated in advance and stored using sodium citrate as an anticoagulant. He developed the first blood bank.

As transfusions became increasingly common after the war, doctors still encountered occasional difficulties. In 1939 Landsteiner with three other scientists identified the Rh factor that caused the transfusion problems. Eighty-five percent of the population have the Rh factor and are referred to as Rh positive. The remaining fifteen percent are Rh negative. If Rh-negative blood is transfused into a Rh-positive individual, a serious reaction can occur.

The war saw the debut of the portable x-ray machine. Radium discoverer and Nobel prize winner Marie Curie organized a campaign to turn cars into x-ray vans to radiograph wounds on the front line. This allowed surgeons to save lives and prevent disability by detecting broken bones, shrapnel, and bullets buried in the flesh.

While many of the medical advances during the conflict dealt with healing the body, there were also psychological wounds that left many soldiers with the uncontrollable tremors of "shell shock." This was known as "soldier's heart" during the Civil War and "combat fatigue" in World War II. Shell shock was the forerunner of today's post-traumatic stress disorder.

The war catapulted medicine forward. In the half-century be-fore the Treaty of Versailles that ended the conflict, doctors had slowly assimilated the bedrock medical concepts of an-esthesia, germ theory, antisepsis, microbiology and pathology. These foundations were the roots of modern medical science. Unfortunately, bedside clinical medicine immediately after-wards still remained rudimentary in many respects. Health-care nonetheless was on the cusp of another leap forward. New standards were established. The cadre of talented doc-tors that returned after the fighting stopped would help med-icine ascend in the years ahead.

Psychiatry

The many soldiers that came back home with "shell shock" after the World War I, serves as a starting point to highlight and review the changes and perceptions of psychiatry over the centuries to the present day. As one author said, "mad-ness and those that treat it are enigmas."

Mental illnesses are unique because they have always been defined by the assessment of the patient's symptoms or behaviors rather than by physical, chemical, or anatomical tests. The word psychiatry is relatively new. It was coined in the 19ᵗʰ century from the Greek words meaning soul or mind and healer. But madness itself is not new—it has been recog-nized since earliest times.

In the 16th century persons with unusual behavior were thought to have let down their moral guard. This resulted in treatments that resembled persecutions. In that era the care of the mentally disturbed was the responsibility of each individual community. Some villages in northern Europe are said to have hired sailors to remove the unruly—hence the origin of the phrase ship of fools. It is an early metaphor of the human condition of those who were mentally imbalanced.

By the 18th century mental illness became incorporated into the medical model and therapy needed to be prescribed by physicians. This produced an increased number of 'mad doctors' who emerged on the scene. They were called alienists. At the same time asylums that were originally meant to be safe places for seclusion, care and restoration came to the forefront.

These institutions tragically morphed into horrifying places of incarnation ostensibly to protect society from criminals, beggars, the poor, prostitutes, and other mad people. Reforms in the 19th century followed when scandalous revelations occurred after investigations of these madhouses. While there were some improvements, those asylums that remained were the last resort for hopeless cases. By the end of the 19th century psychiatry was losing professional credibility. Anesthesia, antisepsis, and other medical advances had fostered intervention for the human problems in the patients of general physicians and surgeons. Psychiatrist on the other hand had yet to make equivalent discoveries that could explain, predict, and cure mental disorders.

Some progress was made in the beginning of the 20th century when Sigmund Fraud (1856-1939) and his followers promulgated psychoanalysis as one answer to certain mental disturbances. It was successful in some cases, but in most it did not curtail the demons within the mentally disturbed.

In an effort to control those with depression and schizophrenia more radical violent treatments were tried. These included insulin shock, electroshock therapy and surgical prefrontal lobotomies in the 1930s and 1940s. In many cases these were desperate measures of well-meaning psychiatrists to do something for the masses of forgotten patients in mental hospitals.

In the post-World War II years, a new generation of psychiatrists came to the forefront. They promoted psychodynamic and psychoanalytic concepts that emphasized the social factors in mental health. They also believed in the early diagnosis of mental disorders and subsequent treatment in community health centers in towns and cities throughout the country. In the United States this concept received the blessing of the federal government. Congress' grants to the states in the 1950s, promoted psychiatric and psychological services.

At the same time medical scientists were laying the basis for the emergence of antipsychotic drugs. This led to the introduction of Librium and Valium in the 1950s. Chemotherapy became the preferred treatment for schizophrenia, manic-depression, and other diseases in the mentally ill. The effect was to dramatically halt the use of prefrontal lobotomies and to minimize and change electroshock therapy techniques.

The length of stay in mental hospitals decreased as there was an increase in public awareness in the inhumane conditions in many psychiatric facilities. The Community Mental Health Act in 1963 was a game changer in the United States. Mental illness became less stigmatized as treatment moved from the inpatient to the outpatient setting. There was deinstitutionalization of the mentally ill. The result was a mass exodus of patients from mental institutions. The cities and towns where they went to live were not ready for the influx and chaos that these unstable individuals frequently brought to their communities.

The experts had not anticipated this downside to this new approach that emphasized prevention and cure outside the hospitals setting. Psychiatrists and social activists overlooked the fact that thousands of chronically mental ill were incapable of functioning on their own. Communities into which mental institutions discharged their patients were ill prepared to receive them. The hospitalized mental population in America dramatically dropped from 559,000 in 1955 to 138,000 in 1980. By 2000 the institutionalized population had declined to less than 10% of what is was just 50 years earlier.

No provisions were made to coordinate the various federal, state, and local mental health facilities. Regional authorities were often unwilling or lacked funds to help. The result was that many of these former patients who were incapable of coping in society were simply left in the street. Early on a few places tried to help these previously institutionalized people, but the majority were just shown how to obtain welfare and then left to sink or swim. Today as many as one-third of the homeless in America are severally mentally ill. The more fortunate ones are periodically arrested and sent to mental facilities only to be returned to the street until their conduct again becomes too outrageous. A 2017 United States Justice Department report estimates that 26% of prison inmates suffer from a psychosis.

In the 1990s Zolof, Prozac, and Paxil came to the market place. In a sense these were designer drugs developed over three decades of growing understanding about the role that the chemical substances dopamine and serotonin play in the brain. Pharmaceuticals are now the mainstay for many minor mental disturbances. They are also used for psychosis and as adjuvants to psychotherapy for other problems. Some psychiatrists now believe that their job is to provide biological pharmacological methods while leaving psychic methods for psychologists. There is still a long way to go in the care of the mentally ill.

The Influenza Pandemic of 1918-1919

In the 21st century concerns for epidemic diseases like Ebola is nothing compared to the influenza pandemic of 1918-1919. It was a worldwide outbreak of frightening virulence. The statistics are hard to believe. World War I claimed an estimated sixteen million lives. The global death toll from the 'Spanish Flu' as the pandemic was called has been estimated at anything between 50-100 million people. That is more than any other universal epidemic outbreak in recorded history.

Greater than 25% of the United States population became sick, and some 675,000 Americans died during the pandemic. Few locations in America were immune to this deadly disease. It involved major cities to remote Alaskan communities. In most locations it crept onto the scene silently leaving a trail of death in its wake. It took less than a month to spread across the entire United States.

The outbreak came in three separate waves throughout 1918 and 1919. Those that were afflicted had high fevers and severe muscle and joint pain from which they recovered and generally had immunity. But up to 10% of these patients affected developed massive pneumonia that caused their demise.

It was first observed in Europe and rapidly came to the United States and broadened around the world. It was a highly contagious virus that spread easily. It was unusual in that it struck down many previously healthy young people who normally would have been resistant to this type of infection. When it occurred, doctors and scientists were unsure of its etiology or how to treat it. Unlike today, there were no effective vaccines or antiviral drugs at that time. It was not until in the 1930s and 1940s that the different viruses that caused the massive illness were isolated.

As the disease widened, schools and businesses emptied. Telephone and telegraph services ceased as operators called in sick. Funeral parlors ran out of caskets, and bodies went uncollected in morgues. Hundreds of thousands became orphaned and widowed. It caused a severe disruption to the economy.

It faded almost as fast as it started with the third wave that had a less virulent mutant virus. In the years following 1919, Americans and other nationalities seemed eager to forget the pandemic, and even today it does not gain much attention in the history books.

Public Health Advances in the United States

The influenza epidemic highlights an often forgotten aspect of healthcare—the role the community at large has in promoting the health of its citizens. For the first three centuries of American, public health laws fell into three main categories:

- Sanitary Laws: These were a mixture of esthetics and a desire to promote health. Early on most public officials believed that dirt and crowding were conductive to sickness. It was clear that garbage, dead animals and over flowing privies were offensive to the senses.

- Quarantine Laws: Well before the bacteriological revolution in the mid 19th century, many were convinced that epidemics were spread from person to person and that isolating the sick could reduce some disease. This resulted in a series of quarantine measures.

- Finally, laws relating to food supply included town ordinances that were enacted to promote better sanitary conditions in public markets.

These three areas received scant attention from the medical profession until the mid 19th century. Then doctors were beginning to call attention to the deplorable living conditions of workers and a need for sanitary reforms. The Civil War delayed this movement. By 1870 the sanitary movement regained its steam. New York City created the country's first permanent board of health that became a model for others to follow. By the 1880s health officials were beginning to widen their concern to include urban housing, air pollution, the health of children, and vital statistics.

In the last decades of the 19th century the wealthy generally were moving to the outskirts of the cities leaving behind their former residences to be occupied by an influx of people seeking employment. Squalid circumstances soon developed. In one small area in Chicago in the 1890s there were sleeping rooms without outside windows and less than 3% of families had their own bathrooms. The situation was compounded by inefficiency and corruption. Political patronage for public sanitation jobs of street cleaning and garbage removal saw not much work being done. These conditions led to a high mortality rate among children and infants in the large cities.

In rural areas and small towns health conditions were better for the lower income groups than in the cities. The roomier environment could better absorb the limited quantities of garbage and waste. Moreover, the water supply was usually cleaner, and food provisions were fresher and more plentiful.

By the late 1890s medicine was beginning to make a difference. Physicians had learned to make more accurate assessments. They identified sources of infection and the mode of transmission. Promoting knowledge of personal hygiene became a part of the process. Medicine was entering into an era to improve the effectiveness of public health. This was never more evident than with diphtheria and tetanus. Here the introduction of antitoxins resulted in a rapid decline in mortality and a general rise of life expectancy in this period.

The 20th century brought modifications that were often led by physicians with political influence. A shift from local and state community efforts at public health to the federal government particularly in the areas of child welfare started in 1921. Even with this federal awareness, progress was slow going until after World War II. Then a massive infusion of money came.

Out of this new funding one of the most significant accomplishments in 20th century America society was to bring better hygienic living within the reach of a large portion of its citizens. By the 1950s safe running water, bathrooms and kitchens with good plumbing, regular sanitary waste disposal facilities, and central heating units had become normal expectations for most people in their housing. At the same time, the ideas for good personal hygiene—cleanliness, toothbrushing, proper diet, and exercise—more than ever had come to be taken for granted as standard parts of a social ethic and individual responsibility.

Paralleling these efforts there was another public health issue—nutrition. Most post war physicians like their predecessors were busy treating aliments and ignored dealing with the dietary practices of their patients. This allowed outside groups like herbalists, holistic practitioners and other alternative medicine groups to step into the void. A proliferation of health food stores came into American cities and suburbs. This has led to a greater awareness and the need for public health inspection and labeling of food products. In the 21st century the community's public health continues to be a central issue in America and throughout the world with ongoing changes occurring in many areas .

The Interwar Years

Between the end of World War I and World War II, the medical ideas and assumptions already in place largely remained intact and were built upon rather than being jettisoned or radically modified. This was evident across the board in all aspects of healthcare.

Orthodox medicine consolidated its authority. Universities and their associated hospitals were increasingly made into the locations of medical education and where the standards of care were set. Hospitals grew in size and number and business like management increased. Specialization within the medical profession became more evident. Technologies used in diagnosis and treatment grew in variety and size. The pharmaceutical companies became larger and introduced new forms of organization, management, and marketing.

All these various worldwide changes were often seen as evidence of modernization or modernism. With this new vital force there was a slow shift of the discovery and implementation of new ideas from Europe to America. By the end of World War II, the United States would become the dominant factor in nearly all aspects of medicine. Before that happened several intervening events that made medical history took place.

The Discovery of Insulin

One of the greatest modern public health issues is a disease that has been apparent since the first century A.D. in Greece—diabetes. Its symptoms have been recognized since ancient times. In the 17th century English physicians determined whether patients had diabetes by sampling their urine. If it had a sweet taste, the diagnosis was made.

This method of monitoring for the disease and blood sugar went largely unchanged for centuries. Before the invention of urine strips in the 1960s, patients played chemist with a collection of test tubes to see if sugar was present in the urine. In the early 1970s the first portable glucose meter appeared. It initially weighed over three pounds. Today they are quite small and accurate.

With the cause unknown, ancient treatments were not effective. The Greeks recommended oil of roses, dates, raw quinces and gruel. The 17th century doctors prescribed jelly of viper's flesh, broken red coral, sweet almonds, and fresh flowers of blind nettles. There were no real cures until the discovery of insulin.

In 1921, a benchmark medical event in Canada took place. Physicians Frederick Banting and his assistant Charles Best discovered insulin in the pancreas. Their first patient was a young boy dying of diabetes. After their injection of a refined extract of insulin, his dangerously high blood sugar level dropped to near normal within 24 hours. His life had been saved. The news of this dramatic event spread like wildfire across the world.

Shortly thereafter the two different kinds of diabetes, Type I (insulin sensitive) and Type II (insulin insensitive) were identified. Additional research created better types of insulin that had a longer duration of action. In the 1950s oral medications became available for patients with Type II disease. The development of the insulin pump in the 1970s and 1980s that dispensed a continuous insulin dosage through a small cannula was a significant improvement. Its actions mimicked the body's normal release of insulin. The researchers in the 21st century continue to explore for a solution that goes beyond just treating the effects of the disease.

Penicillin and Antibiotics

The discovery of penicillin by Scottish biologist and pharmacologist Alexander Fleming (1881-1955) reshaped the world of clinical practice of medicine forever. His drive to defeat infections was because of his experience treating battlefield wounds in World War I. As a young doctor he observed thousands of soldiers die from tetanus, blood poisoning and gangrene. Fleming soon realized that the aseptic methods championed by Lister worked reasonably well in civilian hospitals but failed in wartime. The chemical antiseptics were more dangerous to exposed human battle wounds than the potential for invading bacteria.

After the conflict he began his own research on anti-bacterial substances. According to what could be called the penicillin myth a mold spore supposedly drifted through an open window into Fleming's laboratory. It settled on a petri dish on which he was growing staphylococci bacteria. While at that time contamination of bacteriological materials with molds was a common laboratory accident, it generally was considered a sign of poor technique.

Fleming later remarked that he would have made no discoveries if his laboratory bench was neat and tidy. His polluted plate was left among stacks of dirty petri dishes when Fleming went on vacation. On his return in August 1928, while talking with a colleague, he noticed out of the corner of his eye that one petri dish with the staphylococcus bacteria close to the edge where mold was present had dissolved.

He pursued the chance discovery of the mold produced substance. He called it penicillin. The material destroyed not just the staphylococci but also streptococci, gonococci, meningococci, and pneumococci—these were the most harmful bacteria known at the time. Further research showed penicillin had no toxic effect on healthy tissue nor did it impede white cell defense functions. However, it was both hard to produce and highly unstable. Therefore, it was considered clinically unpromising by contemporary scientists. A discouraged Fleming as a result did nothing. The medical community gave the 1929 publication of his research little notice. Penicillin remained a laboratory curiosity for a decade.

It took the approach of another war in 1938 to get others interested in studying natural antibacterial substances. This directed two Oxford researchers—Australian pathologist Howard Florey (1898-1968) and chemist Ernest Chain (1906-1979) who fled Nazi Germany—to Fleming's original article on penicillin. Two years later in 1940, their treatment on streptococci infected mice was a success. The following year they gave penicillin to the first human—a 43-year-old policeman who had contracted a mixed infection of staphylococci and streptococci. He rallied almost immediately, but unfortunately there was not sufficient penicillin available to save his life.

With resources strained by the ongoing war, British pharmaceutical companies could not develop a new drug forcing Florey and Chain to go to the United States. Because research on penicillin was closely associated with America's military needs, drug companies immediately became involved with the financial support of the government. They soon were producing penicillin, and major clinical trials were quickly under way. By the time of the Normandy invasion in June 1944, enough penicillin had been manufactured to treat all severe causalities among the allied troops. Military doctors no longer had to look on hopelessly as wounded soldiers died from unstoppable infections.

Believing that it should be made freely obtainable to the world, the two British researchers never attempted to patent penicillin. Fleming also fought to make sure the word "penicillin" would not become a commercial trademark. It was truly a wonder drug. By 1950 it was being abundantly produced and became readily procurable to the civilian population where its effects were dramatic. It powered a revolution in medicine.

Fleming's keen eye on that petri dish and the determined minds of Chain and Florey brought the miraculous "mold juice" into the world ushering in the era of antibiotics. The three received the Nobel Prize in Medicine in 1945 for their work.

While penicillin became the first wonder drug, the actual initial breakthrough in the fight against infections occurred in

1932. A German bacteriologist noted the effects of the red dye Prontosil on streptococcus infections in mice. Its active agent was sulfanilamide. French, American, and British scientists subsequently created a host of new sulfa drugs. It is credited with saving the lives of many including Winston Churchill and was used during World War II on wounds and to fight gastrointestinal infections. While this medication was a marvelous discovery, it had definite limitations and numerous side effects. Its use declined with the mass production of penicillin. Sulfa drugs continue to be prescribed today for conditions like urinary tract infections and acne.

After the mass development of penicillin in the United States, American scientists moved to the forefront in the investigation for additional products capable of destroying pathogenic organisms in man. The years during and after World War II opened the era that unfolded many new antibiotics. American drug companies, cut off from Europe, were forced to step up their own research. In 1943 investigators at Rutgers University developed a new drug called streptomycin that was effective against tuberculosis. In the 1950s two American and one German pharmaceutical concern simultaneously discovered another antitubercular drug. Since then at least five more products for tuberculosis have become available that cure most cases.

A large family of antibiotics used to treat a broad spectrum of illnesses became accessible relatively rapidly by the mid 20th century. Tetracycline was one of the first. Patented in 1950 it initially played an important role in stamping out cholera.

In 1957 Nystatin became available to treat patients at risk for fungal infections. Two employees in the New York State Department of Health developed it from bacteria and named it after where they worked. Thus its appellation became "N," "Y," "S," and "T"—Nystatin. The two scientists donated all the royalties from their invention to a non-profit group for the advancement of academic scientific study.

Not long after the introduction of penicillin it was discovered that resistant strains of bacteria produced an enzyme called penicillinase that inactivated penicillin's effect. Scientists modified the nucleus of penicillin to produce a number of synthetic variants resistant to penicillinase.

When antibiotics were first introduced, no one predicted that the overuse of these medications would cause the emergence of stronger antibiotic-resistant bacteria. The mutations in these infectious agents pose an enormous danger for the future. Doctors are learning to curb the number of antibiotics prescribed and to use them only when no other remedy will help. But patients often insist on receiving inappropriate antibiotic prescriptions from their healthcare providers for almost every inflammatory illness. This practice inadvertently helps promote the growth of superbugs.

Pernicious Anemia

The year 1934 was pivotal on the United States' road to becoming a world authority in medicine. That year is when

America gained its first Nobel Prize in Physiology and Medicine. It was awarded to American scientists for discovering a cure for pernicious anemia. How this happened is another chapter in the expansion of medical knowledge in the 20th century.

After World War I some scientists began to shift their attention from the bacteriological external agents of infection to other causes of disease. As investigators widened their medical inquiry they found that the cause of ill health was on occasion not the presence of something like bacteria but rather the lack of a substance within the body. Some diseases resulted from a deficiency in a group of organic compounds called vitamins. They are required to be obtained in diet because they cannot be synthesized by humans. By the 1930s American scientists had identified and described the effects of Vitamins A, B1 (thiamine), B2 (riboflavin), B3 (niacin), B9 (folic acid), C, D, and E. It was in this direction that Americans were recognized.

Before the Nobel laureate Americans discovered its cure, pernicious anemia was fatal. It had been first recognized in the mid 19th century. It is now known that it is caused by the impaired absorption of vitamin B12 in the stomach of patients affected with the disorder. There are a multitude of symptoms in those who have vitamin B12 deficiency including severe neurological disorders, gastrointestinal symptoms, and cardiac malfunctions.

Early in the 1920s clinicians began examining blood samples of anemic patients and noted low red blood cell counts. They treated pernicious anemia patients initially with blood transfusions. But this had no effect on the disabling neurological symptoms. After five years of study a team of three American physicians—Drs. George Wipple, George Minot, and William Murphy—determined the simple truth: a dietary deficiency caused pernicious anemia. They demonstrated that it could be overcome through a nutritional regime high in liver. For this they were awarded the Nobel Prize.

The winning research left an important question unanswered: what was stored in the liver that cured pernicious anemia? A young doctor from Harvard—William Castle—who had worked with the Nobel laureates picked up where their research left off. He determined that the stomach secreted a protein—intrinsic factor—that brought about the absorption of the liver substance that the Nobel Prize winners had previously stumbled upon. Then later in 1948 the material from the liver was isolated and called Vitamin B12. Physicians then began treating pernicious anemia with monthly injections of vitamin B12.

The 1934 award cannot be underestimated. It was a tremendous psychological boost to American clinicians and researchers. The United States was no longer a hinterland in medicine. It validated the nation's medical authority internationally like nothing before.

Changes in Medical Education

As medicine became more sophisticated so did the training of physicians. The Flexner report in the 20th century's first decade had set the pattern of medical education both in and outside America from which better physicians and research scientists emerged. The developments in medicine and society forced ongoing changes in the medical curriculum to include more instruction and clinical experiences in obstetrics, psychiatry, neurology, pharmacology, public health and preventive medicine.

With the various modifications that required greater knowledge by individual doctors, it was evident that a single physician could not be an expert in all areas. The general practitioner's role was being slowly pushed aside. Medical specialization became the logical outcome of modern scientific and medical knowledge. Graduate training from radiology to endocrinology that started in the pre-World War II era began the transition. The ubiquitous presence of the physician generalist slowly declined. Specialties helped spawn new societies, journals and certification boards. The generalist was initially left on the sidelines. Family physicians eventually organized themselves much like the specialists and started their own residencies and certification boards.

This evolution in medical education and specialization has overall produced a better qualified cadre of medical professionals. For the most part this has meant improved treatment with the increased application of sophisticated diagnostic and therapeutic tools by experienced physicians at large multi complex hospitals. As one author noted "while doctors in hospitals were once feared as a place to die because so little could be done to avert death, some people now fear hospitals as places to die because so much can be done."

Medical Advances in World War II

A well-known medical historian observed that "one of greatest ironies of warfare throughout modern history is that the more brutal and widespread the conflict, the greater are the resulting medical advances."

From a medical standpoint World War II remains one of the most successful conflicts that armed forces have ever fought. Accomplishments spanned the entire spectrum of medicine. The increased medical research in the United States after World War I gave the American military and its allies a distinct advantage over their adversaries. In the European theater just 4% of injured American soldiers died during World War II, contrasted with 10% of enemy combatants.

Advances were met in a number of areas. These included:
- Pioneer blood work in dried plasma combined with the nationwide donor effort permitted more effective management of life threatening injuries and made transfusions commonplace.
- Physicians strictly applied public sanitation measures for the disposal of human waste. This curbed a wide variety of illnesses that had arisen in earlier wars.

- Newly developed medicines like penicillin and vaccines combated wound infections, malaria and yellow fever.
- Surgeons skilled in orthopedics preserved limbs that in previous conflicts would have been lost. Similarly, technical advances in operations on the abdomen and chest saved countless lives.
- Psychiatrists diagnosed and successfully treated stress related disorders.
- Ancillary medical workers used novel medical evacuations procedures on the battlefields and innovative organized acute care hospitals.

Like in prior wars, World War II gave physicians intense practical experiences. Unlike previous conflicts it arrived at a moment in history when American medicine was in ascendancy over its European counterparts. There was a newfound faith in American doctors by the government. Federal authorities actively embrace this positive development during the war.

This was no more evident than in the ongoing struggle between the generalists and specialists. The United States military recognized that doctors with advanced levels of training and special knowledge were key to successful medical outcomes. They awarded higher ranks to board certified medical specialists. This de facto recognition of the process of board certification exams effectively ensured that it would become a mainstay of American medicine after the war and an integral part of the nation's healthcare delivery system in the future.

During the conflict the federal government took the lead and organized groups of medical researchers and directed their activities toward a common good. This effort persisted after the combat concluded with increased funding of civilian investigations.

With peace in 1945 American medicine was at the starting line of an advanced era of medical diagnosis, treatment, and therapeutics. Driven by a dynamic research ethos and by a newly minted conviction of therapeutic relevance the scale of the medical enterprise would exponentially explode into the 21st century. With improved methods of mass communications, the populace more than ever could now personally participate by learning about new cures and the benefits that medicine could bring to them. This helped accelerate the process of innovation as opportunities for commercial success became apparent.

CHAPTER TWENTY
Medical and Surgical Advances
in the Second Half of the 20th Century

Cure for Polio

The post war funding of medical research for specific diseases is best illustrated by the search for a cure of polio or infantile paralysis. This disease had a devastating effect on children. Polio had been present for a long time. This viral based disease spread through fecal matter to the bloodstream and ultimately attacked the nervous system.

Children who had lived in the 19th century and earlier when sanitation standards were lower became immune to the infirmity through continuous exposure. After the 20th century improvements in sanitation fewer children were exposed to the contagion, but those that became fell ill encountered a more virulent form of the disorder. In 1952, the United States had the worst polio epidemic in the nation's history. Of the nearly 58,000 cases reported that year, 3,145 died and 21,269 were left with mild to disabling paralysis.

Much of the research into polio was undoubtedly due to the crippling of President Franklin Roosevelt and the creation of the National Institute for Infantile Paralysis. Further as the public's involvement with the "March of Dimes" campaigns increased it accentuated the need for a cure.

In 1949 the technique of cultivating the poliovirus in non-neural tissue resulted in the separation of the three major strains of polio. This led scientist teams with different approaches to the challenge into a race to produce the first vaccine. One group headed by Jonas Salk aimed to employ an inactivated or dead virus in a vaccine. Another faction headed by Albert Sabin believed that working with a live virus was more effective and less dangerous than handling a dead one. He explored a vaccine that used a live weaken form of the virus.

The first to succeed was Salk. His vaccine from an attenuated dead virus was ready in 1953 for large-scale clinical trials. Unfortunately, there was an initial set back due to one faulty lot of the vaccine, but by 1955 Americans knew that Salk's vaccine was a success. When it was made available, it was given by injection.

Salk's masterstroke did not deter Sabin from continuing to work with the live virus that as a vaccine could be orally administered. In the late 1950s, Sabin conducted successful trial runs first on monkeys and then on humans—including his own family. In 1960 the US Public Health Service approved it for manufacture.

The extreme professional competition between these two scientists developed into a personal feud. Salk and Sabin never made peace with each other on their two different views of the polio vaccine. While both vaccines were effective, Sabin's grew to be more popular because his could be given orally instead of by injection making it more acceptable. Though Salk is more commonly associated with developing the vaccine—and he certainly did so first—Sabin's has proven to be more beneficial to the world in eradicating the disease.

Oral Contraceptive

The latter half of the 20th century saw an influx of medications beyond antibiotics. Everything from hypertension and atherosclerosis to dermatological disorders had pharmaceutical companies inundating the medical profession with promising new cures. Few medications created the stir that the oral contraceptive pill did at its introduction in 1960. It changed the lives of women by permitting them to personally control the number of children they wanted, something that had been difficult in earlier times.

The work of several scientists contributed to the ultimate development of an oral contraceptive. Gregory Pincus, who was initially a Harvard, assembled a team outside of the university setting. They demonstrated that the hormone progesterone prohibited ovulation. The group then developed a method to create hormones from natural substances. This led to an oral form of progesterone.

In Puerto Rico a series of clinical trials were conducted. Women eagerly sought what became known as the Pill. Seventeen percent of these volunteers suffered side effects including some that were quite debilitating. But these adverse symptoms did not deter the women, the scientists, or even the Federal Drug Administration. In 1957 the agency approved of one of the main ingredients for the treatment severe menstrual cramps. After some modifications it was sanctioned three years later as a contraceptive. Since then many versions of the birth control pill have been produced. The efficacy remains quite high as scientist have found ways to reduce the side effects by creating a medication with the lowest hormone dosage possible.

While researchers made the Pill possible, the social conditions needed for the acceptance of the scientific advancement would not have been in place without Margaret Sanger in the beginning of the 20th century. If Gregory Pincus was the father of the Pill, then no doubt Sanger was its mother. In the pre-World War I period there were laws in the United States that classified contraceptive information and devices as immoral obscenities. Those that attempted to advocate for birth control during this time were threatened with imprisonment.

Sanger began to challenge these statutes . She came from a large poor Catholic family where she watched her mother suffer with 18 pregnancies, seven of which ended in miscarriages. After putting herself through nursing school, she was thrust into a world in which only doctors could talk about birth control, and nurses were not permitted to comment.

She became an activist and started a newspaper in 1914 called "The Woman Rebel." She coined the term "birth control" for the first time. As time passed Sanger started birth control clinics and eventually created the Planned Parenthood Foundation. She campaigned tirelessly to give women control of their reproductive lives. Her agitating confrontations led to being arrested on a number of occasions.

In the process some wealthy women came forward to help in funding research for an oral form of birth control. With money in hand, Sanger approached Pincus in 1951 to lead the research on finding the perfect contraceptive bill. She died in 1966 having seen her lifelong work come to fruition.

Surgical Advances:
Heart Operations, Kidney Transplants and Laparoscopy

Like other areas of medicine surgery underwent a transformation in the post-World War II years. Multiple refinements advanced the field. Technological improvements like operating microscopes, laparoscopes, and implants have benefited all surgical subspecialties. There are 3 areas that represent the most dramatic public face of surgery in the late 20th century: cardiac operations, transplant surgery and laparoscopy.

Before World War II heart wounds were treated conservatively. During the 19th century a few adventurous doctors experimented with performing repairs to the pericardial sac surrounding the heart. In the beginning of the 20th century the first successful operation on the heart was carried out. A German surgeon repaired a stab wound to a patient's right ventricle.

In World War II American doctor Dwight Harken was stationed near the front lines in Europe. Faced with seemingly fatal cardiac wounds, he began to remove bullets and shrapnel from the heart after experimenting on animals. He created a novel technique where he opened a small hole in the cardiac wall and inserted his finger to remove the foreign element.

Despite Harken's success, there were numerous problems encountered particularly if surgeons desired to correct congenital heart defects and abnormalities of the heart valves. Physicians needed to be able to work inside the heart without their patients bleeding to death. As a consequence, only limited progress could be made until a circulatory support system was designed.

Dr. C. Walton Lillehei (1918-1999) at the University of Minnesota pioneered the initial work on an 11-year-old child with a heart defect. In this early case, Lillehei used the patient's father as a heart lung machine. The boy's dad was anesthetized next to his son. The boy's blood was routed through the father's circulatory system where it could be oxygenated before returning to the boy's body via the carotid artery. This innovation gave Lillehi 19 minutes to make the intra-cardiac repair. With this technological improvement Lillehei led the way for open-heart surgery and is often called the father of that procedure.

While Lillehi's technique enhanced the chance of success further advances were impossible until a more reliable method of circulatory support was devised. This came from John Gibbons of Jefferson Medical College in Philadelphia who spent over 20 years experimenting with various devices. His effort was rewarded in 1953 when his heart-lung machine made it possible to safely perform open-heart surgery on humans. By 1960 the use of the instrument was common. This along with improved anesthetic procedures facilitated the work of the cardiac surgeon.

Meanwhile in Texas Drs. Michael DeBakey and Denton Cooley developed novel methods to deal with aortic tears and aneurysms with their introduction of grafts in 1964. Elsewhere in that same year another major development in cardiac surgery occurred. This was the coronary bypass. In this operation sections of a patient's own veins or arteries were harvested and used to circumvent a blockage in one or more of the heart's coronary arteries.

In December 1967 the first human-to-human heart transplant was performed by Dr. Christiaan Bernard (1922-2001) in Cape Town, South Africa. This was followed by similar surgery at Stanford University a month later. After that a plethora of these operations around the world were done. Unfortunately there was no long-term success with those initial cases.

Today, with many refinements since Dr. Bernard's first patient, close to 85% of heart transplants live at least a year after surgery, and greater than 60% live 5 years or more. Medicare pays for cardiac transplants, but the major problem is the shortage of heart donors.

Cardiac transplantation would not have been possible without the earlier work on kidney transplants. This operation was one of the combined medical-surgical triumphs of the 20th century. The procedure was a testament to the vision and cooperation between surgeon Francis Moore and internist George Thorn at Harvard's Peter Brent Brigham Hospital.

Moore was an atypical surgeon beyond his brilliance in his specialty where he built a worldwide reputation. Unlike the high ego of many surgeons, Moore saw the need for cooperation between different specialties to face modern medical challenges. At his institution he found this collaboration in the ingenious chief of internal medicine, George Thorn. His achievements were well known including developing the first dialysis machine. Together they jointly oversaw the world's first kidney transplant.

In October 1954 Richard Herrick had only a few weeks to
live because of kidney failure secondary to a severe acute
inflammation called glomerulonephritis. His only chance for
survival was a kidney transplant. This surgical procedure had
never been previously successfully performed. While the odds
did not favor Richard, two things made his case unique and
proved lifesaving. First, he was admitted to Brigham Hospital,
home of Moore and Thorn, and second, he had an identical
twin brother.

While little was understood about organ rejection, Moore and
Thorn knew Richard's identical twin status and their indistin-
guishable immune systems meant that Richard's body would
not reject a donated kidney from his brother. After stabilizing
Richard with Thorn's dialysis machine, the doctors had to
convince Richard's brother Ronald to do something that no
human had ever done before: voluntarily allow surgeons to
remove a major organ for the benefit of another person.

There were ethical concerns raised by some but not by
Ronald. He later remarked, "They left it for me to decide and
I was the one who was going to do it; they did not make the
decision for me." He confirmed his decision to proceed while
in his hospital bed the night before the surgery when his
brother Richard passed him a note that said, "Beat it, go
home while you can." Ronald responded, "I am here and
staying here."

The next day December 23, 1954 after removing Ronald's
kidney the surgical team led by Moore secured the kidney in

Richard's lower abdomen. Crystal clear urine was immediately evident. The news of the success of the world's first long-term kidney transplant soon spread.

As for the Herrick brothers they returned to their normal lives. Richard married one of the nurses at the hospital and raised a family. But nine years later glomerulonephritis returned and he died in 1963. Ronald, the world's first organ donor, died at age seventy-nine in 2010 of complications following heart surgery.

The success of the first kidney transplant changed medicine beyond its technical feat in more profound ways. It helped break a psychological and spiritual barrier that viewed the human body as a sacred object able to receive medical care but not designed to provide it. Transplantation was here to stay. With the discovery that cortisone suppressed the immune system and the body's resistance to foreign tissue, this marked the beginning of drug induced immunological tolerance. One did not need to have a twin to receive a kidney. By 1962 kidney transplants were routine procedures.

With surgical techniques on the heart and kidney transplants in the headlines, medical instrument companies were incentivized to search for innovative technologies. The product development teams came up with a number of devices. One that gave doctors a new creative technique was the laparoscope. This led to "minimally invasive" surgery. Through a trocar (a plastic tube appliance) that is inserted into a patient, the laparoscope is introduced into the abdominal cavity. It then provides real-time video images of the patient's inside organs. Surgical tools are inserted and used to perform the operation as the physician views the manipulations on a television screen.

The laparoscope obviates the time-honored tradition of making a lengthy incision. In addition, there is decreased bleeding, fewer infections, less discomfort, and reduced recovery time. With modifications the laparoscope adopted for various different surgeries in the abdomen and pelvis. By 1994 ninety percent of all gallbladder procedures in the United States used laparoscopic techniques.

The biggest prize that laparoscopy claimed was the repair of groin hernias. It is the country's most common abdominal operation. Hernia repairs have been around since the late 19th century when William Halsted at Johns Hopkins described the technique that endured for most of the 20th century until the laparoscope was introduced. He like other surgical professors then and now teach that hernia surgery requires knowledge of the three fundamental elements of any surgical operation:

- An understanding anatomy and pathophysiology
- Familiarity of various repair techniques
- The need for gentle handling of tissues while cutting and separating tissues.

Today, surgeons perform about 40% of all hernia repairs with the laparoscope. While the clinical advantages are very apparent, it comes at a higher cost to the consumer and third party payers.

Atherosclerosis

There are few individuals today that are not aware that atherosclerosis is the progressive and stealth like disease of arteriosclerosis or colloquially called "hardening of the arteries." It begins with a yellowish white swelling called an atheroma. The name is derived from the Greek word *athere*, an expression for a lump of porridge and *oma* meaning growth.

An atheroma consists of microscopic pieces of cellular debris including calcium, cholesterol and fibrous tissue that create an inflammation on a blood vessel's inner wall. It joins with neighboring plaques causing atherosclerosis. The result is a constriction of the inner contours of the artery that reduces and sometimes stops the blood flow. Oxygen to surrounding tissue is decreased in a condition known as ischemia.

Atherosclerosis can occur throughout the body. When the disease affects the heart, the ischemic attack is termed angina. This is frequently accompanied by chest pain, shortness of breath, and sweating. When the tissue of the heart dies from the lack of oxygen, it is produces a myocardial infarction or heart attack.

In the brain, a narrowed or occluded artery causes a variety of symptoms from skin numbness, slurred speech, vision loss, or extremity weakness. If the symptoms are temporary it is a TIA or Transient Ischemic Attack. A permanent more severe case is a stroke with the potential for coma and death. Similarly, atherosclerotic vessels in the legs can cause pain and may if untreated led to infection that fails to heal. Likewise, ischemic changes occur in the kidney and intestines from atherosclerotic arteries.

Atherosclerosis is a most proficient killer. When the annual mortality from both heart disease and stroke caused by atherosclerosis is combined it surpasses death by cancer by 40%. Medical scientists have long advised that the precipitating causes of atherosclerosis include cigarettes, diabetes, fatty foods, obesity, lack of exercise and genetics. News stories warn to cut down on high cholesterol foods to reduce one's chances for atherosclerosis. In the 1980s statin drugs were introduced to decrease a person's cholesterol. Despite this, atherosclerosis continues to be an efficient killing machine.

Surgically a coronary bypass procedure can cure the coronary arteries blockage when atherosclerosis is involved. While very successful it is not without complications. Other interventional methods were sought. It took a German innovator Andreas Gruntzig to change the treatment of atherosclerosis. Working initially on diseased coronary arteries he placed a catheter with a balloon on the end inside an atherosclerotic obstruction. Then, inflating the balloon, the plaque was pushed against the wall reestablishing blood flow. It took him three years to develop a spaghetti-sized catheter with a miniature balloon on the tip. Gruntzig introduced his procedure in 1975 at the American Heart Association's national meeting.

Few events in the history of medicine dramatically alter patient care in a single instance. Gruntzig's success was one. He had managed to relieve heart disease without surgery with balloon angioplasty. It evolved to being used outside the heart where there were clotted blood vessels. The procedure was a game changer in the treatment of the coronary arteries. It led to further modifications including today's stents that permanently traverse the diseased blood vessel.

By 1995 the procedure equaled the number of cardiac bypass operations. More recently angioplasties have out numbering bypasses by 4 to 1. There are some selected patients that drugs alone combined with a better life style can relieve symptoms of heart disease. But most cardiologist today believe that angioplasties and newer drug imbedded stents saves lives and can improve the quality of life.

As to the man who started it all, Andreas Gruntzig became an American medical celebrity. Moving to the United States in the early 1980s he settled in Atlanta, Georgia and worked at Emory University. He became very wealth and capitalized on his success with catheters of his own design. He spent his fortune lavishly including a multiengine airplane. He said to associates, "I like flying because it confirms that I have no fear." Unfortunately this claim proved accurate. He died on October 27th, 1985 when his own arrogance convinced him that with only 58 hours in his new aircraft it could safely travel through an approaching hurricane. He crashed with his wife and dog going down off the Georgia coast.

Cancer

The one disease that has become a focal point for nearly all fields of biomedical research is cancer. It is the prototypical modern disease that is one of the oldest on record. In ancient Greece doctors identified its signature malignant swellings. The Romans' described these growths as carcinos or crablike. One physician from antiquity said, "Just as a crab's feet extend from every part of its body so in this disease the veins are distended forming a similar figure."

The disease defies every rule of biologic sensibilities as malignant cells join together in a pact of destruction. They proliferate uncontrollably and metastasize to distant parts of the body along the way eroding blood vessels and destroying vital centers. Through the centuries doctors discovered that the disease had hundreds of different forms and obeyed few rules. Modern medicine divides cancers into 4 broad groups based upon its cellular derivation.

- Carcinomas arise in cells that cover the surface of the body and line organs and glands. They account for over 80% of all cancer cases including common forms of bladder, breast, colon, lung, pancreas, prostate and skin malignancies.
- Sarcomas are less frequent and originate in supportive tissues like bones, cartilage, ligaments, muscles, tendons as well as blood vessels, and the brain.
- Leukemia's are derived from the blood forming cells of the bone marrow.

- And finally, lymphomas develop in the lymph nodes.

While cancer has aroused dread throughout history, little was known about its etiology. In 1910 the cause of a tumor in chickens was discovered to be a virus. But these findings and reports of similar tumors in rabbits were not fully appreciated until the field of virology was stimulated by polio research in the 1940' and 1950s.

By 1933 cancer had become the second leading cause of death in the United States and had replace tuberculosis as the most dreaded disease. As public concern mounted in the post-World War II years private and public money began seeding cancer investigations on multiple fronts. As the years passed and the search for a cause broadened to immunologists and geneticists, a number of different factors were identified as contributing to the rising incidence of malignancies. In 1971 President Nixon championed cancer research noting that the same kind of effort that took man to the moon should be turned towards conquering this dreaded disease. The National Cancer Act of 1971 generously supported clinical research. It created a network of cancer care institutions and established clinical trials as a model of modern medical care.

The treatment of cancer changed in the 20th century. For most of medical history cancer was within the exclusive province of the surgeon. They frequently removed superficial tumors visible on the body's surface. This provided some relief to patients from pain and the stench of ulcerating sores.

For many years the scalpel could not battle malignancies inside the body. As improved methods for controlling shock, blood replacement and antibiotics were developed it permitted more radical and safer surgery.

At the beginning of the 20th century clinicians began using radiation as a supplemental treatment for cancer. Radiation could reduce the size of tumors preoperatively. The combination of surgery and radium or external beam therapy was the most effective therapy available for the first half of the 20th century. Since then radiation oncologists have improved their techniques to reduce side effects so that radiation is used today for the primary treatment for some forms of malignancy as well as metastatic disease.

In 1947 when Harvard pathologist Sidney Farber discovered that chemical agents could inhibit malignant diseases, cancer chemotherapy soon became a household phrase. It was a new weapon in the fight against cancer. Newer unique types of chemotherapy are now tailored to specific kinds of cancer. The most recent advances in immunotherapy demonstrates significant potential. One form is tumor infiltrating lymphocytes or TILS. These TILS cells are taken from the malignancy of a patient, grown and altered in tissue cultures and then given back to the patient. These modified cells then home in on the tumor and destroy it. This treatment is truly personalized to a particular patient. The future looks promising.

Unlike just 50 years ago a large team consisting of oncologists, radiologists, surgeons, pathologists, and other professionals aid the cancer patient. Much still needs to be accomplished before the threat of cancer can be reduced or removed. Despite the steady progress, the present methods of treatment are quite heroic. One wonders whether future generations of physicians may view current cancer therapies with the same horror that today's doctors look upon the bloody and painful surgery of years ago.

Evidence Based Medicine

As the 20th century drew to a close, academic medical educators in the 1990s began to incorporate the principles of Evidence Based Medicine into medical school curriculums. This paradigm, which had its origins in a book published in 1972 by Archie Cochrane, sets out a specific course of action that stresses the judicious use of relevant information for medical decision making. Simply put this doctrine suggests, that because medical funding would always be limited, resources should be used to equitably provide for those aspects of health care that have been demonstrated in a properly designed evaluation to be effective. Not expectedly some were initially opposed to this concept.

Many medical educators believed that the time was right to bring clinical decision making closer to its scientific potential. The lack of any formal process for establishing a medical diagnostic template had characterized clinical practice since the time of Hippocrates. While this may seem strange to medical outsiders, it is the reason that the "art" of medicine has always been given so much latitude in the training and the practice of the profession.

A major challenge with medicine in the credo of many is that expert opinion and clinical experience is often flawed when it comes to making the correct diagnosis. Academic scholars believe that if medical care is to improve, health care decisions must rely more upon research based evidence than subjective judgments. Medical school professors are hopeful that by incorporating these ideas into the curriculum for future physicians they would embrace the concept of evidence based medicine when they start practicing.

Evidence Based Medicine requires a new skill set. The doctor takes a patient's clinical problem through an elaborate process. This includes: searching relevant literature, appraising both the validity of clinical evidence and the patients' expectations. While this has become a useful technique, it is not applicable to all clinical situations and individual expertise is still appropriate in the final decision for diagnosis and treatment for the patient.

For some this is "cook book medicine." But those trained in the method know when to use it and whether it is relevant to the patient at hand. Its principles were included in the Accountable Care Act passed in 2009.

The Human Genome Project

In the new millennium one of the more amazing areas in medicine has been the human genome project. Its foundation was in 1953 when American James Watson (1928 -) and British citizen Francis Crick (1916-2004) were able to describe the structure of deoxyribonucleic acid (DNA) that is the main constituent of chromosomes. This discovery set the stage for a revolution in molecular genetics. They revealed that the four base pairs of nucleic acids were linked together in a double helix resembling a twisted ladder. This formed an encryption like the dots and dashes in the Morse code. Subsequent research provided detail on the chemical and biological properties of DNA.

Fast forward to the 1980s Watson helped lobby Congress to create the United States Genome Project. The goal of this enterprise was to map out the exact sequence contained in each of the 24 human chromosomes. It took time to develop a technique to achieve their objective. But by 2003, the 50th anniversary of Watson and Crick's discovery of the DNA's structure, 99% of the human genome had been sequenced.

Why is this important? The human genome is a person's complete set of DNA arranged into twenty-three distinct chromosome pairs—the 24th is the set that determines gender. Each chromosome contains many genes that are the basic physical and functional bits of heredity. One surprise is that scientists have found that humans are a great deal simpler than was imagined at the outset. They have shown that the human genome has only 30,000 genes—one-third the amount originally estimated. Researchers continue to explore on how genes function, produce disease and other abnormalities.

Diagnostically, the door is now open for genetic testing that can reveal the predisposition of an individual to a variety of diseases. In the field of pharmaco-genetics precision medicine treatments and vaccines will likely be made to order for a particular individual with a specific genetic makeup. Genome sequencing data collectively also has the potential to be beneficial in improving overall population health. In addition, there is advanced research studying the genes of patients with mysterious conditions that have previously eluded diagnosis.

Alongside of the scientists making this genetic related progress, others wrestle with the ethical issues that these advances bring. There are multiple difficult questions. The issues are centered on privacy and access to an individual's genetic information and whether there will be discrimination against someone with a predisposition to a particular disease.

EPILOGUE

The story of how medicine has evolved from the era of the cave dwellers to the present millennial generation is as much the art of forgetting as it is of remembrance. It is apparent as one scientist so eloquently observed, "medical progress does not move forward with the even pace of the clock's hand. It jumps unpredictably at its own fitting time of ripeness."

The primitive medical practices had magic, religion, ingenuity and some experimentation. Truth was largely based on the human senses of sight, smell, and touch while modern medicine looks to the laboratory for veracity. Doctors from previous generations needed to persuade their patients of their skills in order for them to accept the harsh treatment they were prescribing, otherwise it would not be a success. A similar situation occurs today for the modern healthcare provider who explains a complicated therapy that their patient is about to undergo.

Notwithstanding the multiple obstacles and challenges, early healers produced all manner of success to help the ill and injured without understanding why they worked. Over time a series of creative stages were set in motion by brave souls. They defied authority and societal subversion to change how healthcare had been practiced for centuries. These iconoclastic practitioners were from diverse backgrounds and education. Because they were all willing to challenge the conventional thinking of their era progress was made. This continues to be true today.

In the 21st century new horizons are visible, but so are unanticipated problems. Advancements in medical science have people living longer. There is an unintended consequence of this success. With longevity, a greater concern for degenerative illnesses that effect the elderly will emerge—like in the past with new victories come new challenges.

In the deluge of new medical discovers that are unveiled almost daily, it easy to overlook and underestimate the ingenuity of those doctors and scientists who came before us. Hopefully this book has made the reader cognizant how past generations vanquished long-standing hurdles to achieve today's modern medicine.

Medicine has progressed enormously from that day 40,000 years ago when an anonymous cave dweller for the first time ever splinted a broken leg to today's 21st century millennial who just received a personal genetic profile indicating a curable cancer will occur in the future. The idea is better stated by the author who wrote, "we can only understand where we are, if we know how far we've come."

REFERENCES

Abrams, Jeanne E., *Revolutionary Medicine* (New York and London, New York University Press, 2013).

Ackerknecht, Erwin H., *A Short History of Medicine* (Baltimore, The Johns Hopkins University Press, 1982).

Adler, Robert E., *Medical Firsts* (Hoboken, John Wiley & Sons Inc., 2004).

Armstrong, David and Armstrong, Elizabeth M., *The Great American Medicine Show* (New York, Simon and Shuster, Inc., 1991).

Barry, John M., *The Great Influenza* (London, Penguin Books Ltd., 2009).

Beller, Susan Provost, *Medical Practices in the Civil War* (Charlotte, VT, published by author, 1992).

Belofsky, Nathan, *Strange Medicine* (New York, Penguin Group, 2013).

Bollet, Alfred Jay, *Civil War Medicine* (Tucson, Galen Press, Ltd., 2002)

Bonner, Thomas Neville, *Becoming a Physician* (Baltimore, The Johns Hopkins Press, 2000).

Buchman, Dian Dincin, *Ancient Healing Secrets* (New York, Black Dog & Leventhal Publishers, 2005).

Bryan, William, *The History of Medicine* (Oxford, Oxford University Press, 2008).

Bynum, W.F., *Science and the Practice of Medicine in the Nineteenth Century* (Cambridge, Cambridge University Press, 1994).

Bynum, W.F., Hardy, Anne, Jacyna, Stephen, Lawrence, Christopher, and Tansey, E.M., *The Western Medical Tradition* (Cambridge, Cambridge University Press, 2006).

Carrick, Paul, *Medical Ethics in the Ancient World* (Washington, D.C., Georgetown University Press, 2001).

Cassedy, James H., *Medicine & American Growth 1800-1860* (Madison, The University of Wisconsin Press, 1986).

Cassedy, James H., *Medicine in America* (Baltimore, The Johns Hopkins University Press, 1991).

Dawson, Ian, *The History of Medicine Series: Prehistoric and Egyptian Medicine, Medicine in the Middle Ages, Renaissance Medicine* (New York, Enchanted Lion Books, 2005).

Demaitre, Luke, *Medieval Medicine* (Santa Barbara, ABC-CLIO, LLC, 2013).

Duffin, Jacalyn, *History of Medicine* (Toronto, University of Toronto Press, 2010).

Duffy, John, *From Humors to Medical Science* (Urbana and Chicago, University of Illinois Press, 1993).

French, Robert, *Medicine Before Science* (Cambridge, Cambridge University Press, 2003).

Friedenberg, Zachary, *The Doctor in Colonial America* (Danbury, Rutledge Books, Inc., 1998).

Gabriel, Richard A., *Man and Wound in the Ancient World* (Dulles, Potomac Books, 2012).

Geller, Markham J., *Ancient Babylonian Medicine* (Chichester, John Wiley & Sons Ltd., 2015).

Gonzalez-Crussi, F., *A Short History of Medicine* (New York, Random House, Inc., 2007).

Gordon, Richard, *The Alarming History of Medicine* (New York, St. Martin's Press, 1993).

Hall, Tim, *History of Medicine* (London, Hodder & Stoughton Ltd., 2013).

Kelly, Kate, *The History of Medicine: Early Civilizations, The Scientific Revolution and Medicine, Old World and New, Medicine Becomes a Science* (New York, Facts on File, Inc., 2010).

King, Helen, *Greek and Roman Medicine* (London, British Classical Press, 2001).

King, Lester S., *Transformations in American Medicine* (Baltimore, The Johns Hopkins University Press, 1991).

Lloyd, G. E. R., *Hippocratic Writings* (London, Penguin Books Ltd., 1978).

Loker, Thomas W., *The History and Evolution of Healthcare in America* (Bloomington, iUniverse, 2012).

Ludmerer, Kenneth M., *Time to Heal* (Oxford, Oxford university Press, Inc., 1999).

Magner, Lois M., *A History of Medicine* (New York, Marcel Dekker, Inc., 1992).

Majno, Guido, *The Healing Hand* (Cambridge, Harvard University Press, 1975).

Margo, Curtis E., *Glass Half Full* (Tampa, Clio Mentor, 2009).

Nunn, John F., *Ancient Egyptian Medicine* (Norman, University of Oklahoma Press, 1996).

Nutton, Vivian, *Ancient Medicine* (London, Routledge Taylor & Francis Group, 2004).

Oshinsky, David, *Bellevue* (New York, Doubleday, 2016).

Parker, Steve, *Kill or Cure* (New York, DK Publishing, 2013).

Parker, Steve, *Medicine the Definitive Illustrated History* (New York, DK Publishing, 2016).

Pickover, Clifford A., *The Medical Book* (New York, Sterling Publishing, 2012).

Pomann, Peter E. and Savage-Smith, Emile, *Medieval Islamic Medicine* (Washington, D.C., Georgetown University Press, 2007).

Porter, Roy, *The Greatest Benefit to Mankind* (New York, W.W. Morton & Company, 1997).

Porter, Roy, *Blood and Guts* (New York, W.W. Morton & Company, 2002).

Porter, Roy, *The Cambridge History of Medicine* (Cambridge, Cambridge University Press, 2006).

Porter, Roy, *Time Tables of Medicine* (New York, Black Dog & Leventhal Publishers, 2000).

Range, Caroline, *The History of Medicine in 100 Facts* (Gloucestershire, Amberley Publishing, 2015).

Rosen, William, *Miracle Cure* (New York, Penguin Random House LLC, 2017).

Rosenberg, Charles E., *Our Present Complaint* (Baltimore, The Johns Hopkins University Press, 2007).

Rothstein, William G., *American Physicians in the 19th Century* (Baltimore, Johns Hopkins University Press, 1985).

Rutkow, Ira, *Seeking the Cure* (New York, Simon & Shuster, Inc., 2010).

Starr, Paul, *The Social Transformation of American Medicine* (London, The Perseus Group, 1982).

Steele, Volney, *Bleed, Blister, and Purge* (Missoula, Mountain Press Publishing Company, 2005).

Tannenbaum, Rebecca, *Health and Wellness in Colonial America* (Santa Barbara, ABC-CLIO, LLC. 2012).

Thorwald, Jurgen, *Science and Secrets of Early Medicine* (New York, Horcourt, Brace & World, Inc., 1962).

Toledo-Pereyra, Luis H., *A History of American Medicine from the Colonial Period to the Early Twentieth Century* (Lewiston, The Edwin Mellen Press, 2006).

Shryock, Richard Harrison, *Medicine and Society in America* (Ithaca, Cornell University Press, 1960).

Siraisi, Nancy G., *Medieval & Early Renaissance Medicine* (Chicago, The University of Chicago Press, 1990).

Sournia, Jean-Charles, *The Illustrated History of Medicine* (London, Howard Starke Publishers Limited, 1992).

Walsh, James J., *Old-Time Makers of Medicine* (Radford, Wilder Publications, LLC., 2011).

Weisse, Allen B., *Conversations in Medicine* (New York, New York University Press, 1984).

Wertz, Richard W. and Wertz, Dorothy C., *Lying-In* (New York, Macmillan Publishing Co., 1977).

Zysk, Kenneth G., *Asceticism and Healing in Ancient India* (Delhi, Motilal Banarsidass Publishers, 1991).

INDEX

ABOUT THE AUTHOR

Doctor Stolz is a graduate of The Hill School, Trinity College and Temple University School of Medicine. After completing an internship at Charity Hospital—Tulane University division and a residency in radiology at the Hospital of the University of Pennsylvania, he practiced in Reading, Pennsylvania before retiring and moving to Williamsburg, Virginia.

In retirement, he continues to pursue his long time study of the history of medicine. Since 2008 he has taught courses on different aspects of medical history as a volunteer instructor at the College of William and Mary's Christopher Wren Association. This is an organization dedicated to lifelong learning for seniors in the Williamsburg community. He also periodically publishes medical history related commentaries in Williamsburg's *Virginia Gazette*.

He and his wife enjoy spending time with their four grand-children, two daughters and their daughters' husbands.

CPSIA information can be obtained
at www.ICGtesting.com
Printed in the USA
BVHW03s1910130918
527452BV00001B/7/P